# Studies in Computational Intelligence

Volume 1046



**Series Editor**

Janusz Kacprzyk, Polish Academy of Sciences, Warsaw, Poland

The series "Studies in Computational Intelligence" (SCI) publishes new developments and advances in the various areas of computational intelligence—quickly and with a high quality. The intent is to cover the theory, applications, and design methods of computational intelligence, as embedded in the fields of engineering, computer science, physics and life sciences, as well as the methodologies behind them. The series contains monographs, lecture notes and edited volumes in computational intelligence spanning the areas of neural networks, connectionist systems, genetic algorithms, evolutionary computation, artificial intelligence, cellular automata, self-organizing systems, soft computing, fuzzy systems, and hybrid intelligent systems. Of particular value to both the contributors and the readership are the short publication timeframe and the world-wide distribution, which enable both wide and rapid dissemination of research output.

Indexed by SCOPUS, DBLP, WTI Frankfurt eG, zbMATH, SCImago.

All books published in the series are submitted for consideration in Web of Science.

More information about this series at https://link.springer.com/bookseries/7092

Robert Simon Fong · Peter Tino

# Population-Based Optimization on Riemannian Manifolds

Robert Simon Fong
University of Birmingham
Birmingham, UK

Peter Tino
University of Birmingham
Birmingham, UK

ISSN 1860-949X ISSN 1860-9503 (electronic)
Studies in Computational Intelligence
ISBN 978-3-031-04295-9 ISBN 978-3-031-04293-5 (eBook)
https://doi.org/10.1007/978-3-031-04293-5



This Springer imprint is published by the registered company Springer Nature Switzerland AG
The registered company address is: Gewerbestrasse 11, 6330 Cham, Switzerland

*To Eddie, Irene, William and Ava*
*—Robert Simon Fong*

*To my parents, Maria (Maruška), Palo (Palko), Michal (Miško) ...*
*and laktibrada ...*
*—Peter Tino*

# Contents

# Acronyms

| | |
|---|---|
| CMA-ES | Covariance Matrix Adaptation Evolutionary Strategies |
| EDA | Estimation of Distribution Algorithms |
| Extended RSDFO | Extended Riemannian Stochastic Derivative-Free Optimization |
| PSO | Particle Swarm Optimization |
| RCMA-ES | Riemannian Covariance Matrix Adaptation Evolutionary Strategies |
| RSDFO | Riemannian Stochastic Derivative-Free Optimization |
| RTR | Riemannian Trust Region method |
| SDFO | Stochastic Derivative-Free Optimization |
| TR | Trust Region method |

# Chapter 1
# Introduction

**Abstract** Geometry has a significant role in the advancements of many fields of modern science, and most recently it has been brought under the limelight of information theory and artificial intelligence. Statistical models can be endowed with specialized geometrical structures, and data measurements can be viewed as points on an underlying manifold, a generalization of $n$-dimensional geometrical objects, rather than on Euclidean spaces. This chapter overviews the recent advancements in geometry and optimization theories, and outlines how both the manifold of data and the manifold of statistical models can be combined to construct optimization algorithms from geometrical principles. A detailed book synopsis is included in the chapter to guide the readers through the rest of the chapters.

## 1.1 Motivation

Geometry permeates a wide spectrum of modern science, ranging from the theoretical studies of classification of shape, geometrization of physics, to formal description of rigid motion used extensively in robotics. In the language of modern differential geometry, these objects are studied under the notion of manifold, a formalization of $n$-dimensional geometrical objects.

Advancements of information theory has brought theoretical studies of geometry under the limelight in a variety of applications: notably in fields of physics [SR+10], biology [HKK06, KH05], computer vision [FJLP03, CIDSZ08] and many more, where sets of data measurements can be viewed as points on an underlying manifold rather than being embedded in a much larger Euclidean space. This observation allows us to operate on a lower dimensional manifold, which in turn lowers the complexity of computation, and provides deeper understanding of the intrinsic structure of the underlying set of data points.

Family of probability distributions describing the underlying relations of data points can also be studied under the geometrical setting. The emerging field of Information Geometry [AN00], in particular, studies statistical models endowed with a geometrical structure. This has since been applied to designing computational tools



R. S. Fong and P. Tino, *Population-Based Optimization on Riemannian Manifolds*,
Studies in Computational Intelligence 1046,
https://doi.org/10.1007/978-3-031-04293-5_1

and provide theoretical insights for optimization algorithms and machine learning methods.

The upsurge of interest in geometrical aspects of information theory also brought development in two distinct branches of optimization theories:

1. **[Gradient-based Manifold Optimization]**: The adaptations of "classical" gradient-based optimization methods from Euclidean spaces to Riemannian manifolds [Gab82, AMS09]. The search trajectory, locally described by search direction vectors, can be viewed as a curve embedded on the manifold search space. The decision space[1] is given by a Euclidean space (tangent space) of "search direction" vectors, while the data points belongs to a manifold search space.
2. **[Information Geometrical interpretation of Model-Based Stochastic Optimization on Euclidean spaces]**: The geometry of the evolution of statistical parameters in model-based stochastic optimization over Euclidean space [ANOK12]. This line of work originated from the emerging field of Information Geometry, where the space of finitely parametrized probability densities over Euclidean spaces is viewed as a "statistical manifold". The evolution of statistical parameters in stochastic algorithms can thus be regarded as a gradient-based optimization process on the overarching statistical model manifold.
The decision space is a manifold comprised of probability distribution over the search (solution) space described by the the statistical parameters, whereas the search space is typically a Euclidean space.

Inspired by the geometrical insights of these two branches of optimization theories, this book develops a geometrical framework for stochastic optimization algorithms on Riemannian manifolds that is both theoretically sound and pragmatic. Under this geometrical framework, both the decision space and search space are regarded as manifolds with their own geometrical structures.

## 1.2 Overview

Consider multi-modal,[2] unconstrained black-box optimization problems on Riemannian manifolds $M$:

$$\max_{x \in M} f(x),$$

where $f : M \to \mathbb{R}$ denotes the objective function over the search space Riemannian manifold $M$. The search space manifold $M$ is regarded as an abstract, stand-alone non-linear search space, that is free from an ambient Euclidean space.

[1] Roughly speaking, the **decision space** of an optimization algorithm is a space where the search information is processed and generated.

[2] A real-valued function $f : M \to \mathbb{R}$ is called **multi-modal** if $f$ has more than one optimum (local or global) in $M$.

Population-based optimization methods have been proven effective in solving multi-modal, black-box optimization problems over Euclidean spaces. Notable examples include meta-heuristics such as Particle Swarm Optimization [EK95], and population-based Stochastic Derivative-Free Optimization (SDFO) methods such as Estimation of Distribution Algorithms (EDA) [LL01] and Covariance Matrix Adaptation Evolution Strategy [Han06].

Population-based optimization methods are adapted to tackle optimization problems on Riemannian manifolds in two distinctive fashions, each with their own drawbacks.

In manifold optimization methods developed in recent literature [Gab82, AMS09, AH19], the search space manifold is considered as an abstract, stand-alone non-linear search space. The work in the literature focuses on adapting existing optimization methods from Euclidean spaces to Riemannian manifolds. Examples include gradient-based optimization methods [AMS09], population-based meta-heuristics [BIA10] and model-based stochastic optimization [CFFS10]. Whilst the structure of the pre-adapted algorithm is preserved, the computations and estimations are locally confined by the normal neighbourhoods on the search space manifold. Furthermore, additional assumptions on the search space manifold have to be made to accommodate the adaptation process.

In the classical optimization literature, optimization on Riemannian manifolds falls into the category of constraint optimization methods [AMS09]. Indeed, due to Whitney's embedding theorem [Whi44a, Whi44b], all manifolds can be embedded in a sufficiently large ambient Euclidean space. As traditional optimization techniques are more established and well-studied in Euclidean spaces, one would be more inclined to address optimization problems on Riemannian manifold by first finding an embedding onto the manifold, and then applying familar classical optimization techniques. The search space manifold is therefore considered as a subset of an ambient Euclidean space, which is then described by a set of functional constraints. However, the global structure of Riemannian manifolds is generally difficult to determine, and a set of functional constraints that describes general search space manifolds can often be difficult, or even impossible to obtain. As a result the structure of the underlying manifold search space is often ignored.

The book advances along these two directions. We construct a stochastic optimization method on Riemannian manifolds that overcomes the local restrictions and implicit assumptions of manifold optimization methods in the literature. We address multi-modal, black-box optimization problems on Riemannian manifolds using only the *intrinsic* statistical geometry of the decision space and the *intrinsic* Riemannian geometry of the search space.[3] To motivate and necessitate manifold optimization, we illustrate our approach on a synthetic experiment over Jacob's ladder, a search space manifold that cannot be addressed by classical constraint optimization techniques.

---

[3] That is, the search space manifold does not have to be embedded into an ambient Euclidean space.

To this end, we take the long route and investigate information geometrical structures of statistical models over manifolds. We describe the statistical geometry of locally inherited probability densities on smooth manifolds using a local orientation-preserving diffeomorphic bundle morphism, this generalizes the use of Riemannian exponential map described in both the manifold optimization [CFFS10] and manifold statistics literature [Pen04, Oll93].

To overcome the local restrictions of manifold optimization algorithms and parametric probability distributions on Riemannian manifolds in the literature, we require a family of parametrized probability densities defined beyond the normal neighbourhoods of Riemannian manifolds. We therefore construct a family of mixture densities on totally bounded subsets of $M$ as a mixture of the locally inherited densities. We show that the family of mixture densities has a product statistical manifold structure; this allows us to handle statistical parameter estimations and computations of mixture coefficients and mixture components independently.

This constitutes a geometrical framework for stochastic optimization on Riemannian manifolds by combining the information geometrical structure of the decision space and the Riemannian geometry of the search space, which:

1. Relates the statistical parameters of the statistical manifold decision space and local point estimations on the Riemannian manifold, and
2. Overcomes the local restrictions of manifold optimization algorithms and parametric probability distributions on Riemannian manifolds in the literature.

Using the product statistical geometry of mixture densities, we propose Extended Riemannian Stochastic Derivative-Free Optimization (Extended RSDFO), a population-based stochastic optimization algorithm on Riemannian manifolds, which addresses the local restrictions and implicit assumptions of manifold optimization in the literature.

The geometrical framework also allows us to study the more general properties of Extended RSDFO, previously unavailable to population-based manifold optimization algorithms due to the local restrictions. We discuss the geometry and dynamics of the evolutionary steps of Extended RSDFO using a "regularized inverse" Fisher metric on the simplex of mixture coefficients. We show that expected fitness obtained by Extended RSDFO improves monotonically , and show that Extended RSDFO converges globally eventually in *finitely* many steps on connected compact Riemannian manifolds.

We wrap up our discussions by comparing Extended RSDFO with state-of-the-art manifold optimization methods in the literature, such as Riemannian Trust-Region method [ABG07, AMS09], Riemannian CMA-ES [CFFS10] and Riemannian Particle Swarm Optimization [BIA10, BA10], using optimization problems defined on the $n$-sphere, Grassmannian manifolds, and Jacob's ladder.

Jacob's ladder is a non-compact manifold of countably infinite genus, which cannot be expressed as polynomial constraints and does not have a global representation in an ambient Euclidean space. Optimization problems on Jacob's ladder therefore cannot be addressed by traditional (constraint) optimization techniques on Euclidean spaces, which necessitates the development of manifold optimization algorithms.

## 1.3 Detailed Book Synopsis

This book is organized into two parts. The first part investigates the Information Geometrical structure of statistical models over Riemannian manifolds. This establishes a geometrical framework to construct Extended RSDFO incorporating both the statistical geometry of the decision space and the Riemannian geometry of the search space.

The second part of the book describes Extended RSDFO, a principled population-based meta-algorithm that uses existing manifold optimization algorithms as it's local module. The construction of Extended RSDFO is detailed rigorously from a geometrical perspective, and it's properties derived from first principles. The remainder of the book is outlined as follows.

- **[Part I: Information Geometry of Probability Densities over Riemannian manifolds]**
  In order to establish a geometrical framework for stochastic estimation and optimization on manifolds, we begin by asking the following questions: What is the geometrical structure of globally defined parametrized probability distributions on manifolds? How would the statistical parameters of these distributions reflect the statistical estimations of points on the base manifold search space?
  In other words, we want to study the geometrical structure of families of (finitely) parametrized probability densities supported on the manifold, whose parameters are coherent with the statistical estimations on the search space manifold.[4]
  To answer these questions, we begin by describing the fundamentals of Differential Geometry and Information Geometry.
  In Chap. 2, we establish the domain of discourse of the book and introduce the essential foundations of Differential Geometry: the study of *intrinsic* geometrical structure of manifolds. This provides the foundations for the geometrical framework of both the *search space* and the *decision space*.
  In Chap. 3, we discuss the fundamentals of statistical manifolds studied in the field of Information Geometry. Statistical manifolds are statistical models endowed with a manifold structure. Unlike "classical" Riemannian manifolds, the dual geometrical and statistical nature of statistical manifolds require a geometrical structure that is both *intrinsic* and *invariant under sufficient statistics*.[5] This provides the geometrical framework for the *decision space*.
  In Chap. 4, we survey notions of volume form and intrinsic probability distributions on manifolds in the literature, which can be roughly classified into a "geometrical" approach and a "statistical" approach. The "geometrical" approach focuses on the information geometrical structure of the space of *all* probability measure on

[4] That is, the parametrization of the statistical model can be estimated from manifold data points just as the "classical" Euclidean case.

[5] More formally, the geometrical structure over a statistical model $\{p(x|\theta)\}$ would be invariant under *both* reparametrization of $\theta$ *AND* remapping of $x$ under sufficient statistics of $\theta$.

the base manifolds, whereas the "statistical" approach aims to re-establish point estimation (locally) on manifolds.

In Chap. 5, we begin by discussing how neither the "geometrical" nor "statistical" approach described in Chap. 4 is suitable for our purpose of establishing a geometrical framework for stochastic optimization on manifolds: the "geometrical" approach is too general whereas the "statistical" approach is too restrictive.[6] We therefore combine the *essence* of the two branches and develop the notion of locally inherited probability densities on $M$.

In particular, we construct probability densities on manifolds geometrically as elements of the density bundle whose parameters are coherent with statistical estimations on manifolds. The proposed framework preserves both the information geometrical structure of volume forms and the statistical meanings of point estimations. This generalizes the "statistical" approach using both the insights of the "geometrical" approach and the machineries of information geometry. However, the locality of the "statistical" approach still persists.

In Chap. 6, we extend the notion of parametrized probability densities over manifolds beyond the confines of a single normal neighbourhood, overcoming the locality of the "statistical" approach described in Chap. 4. In particular, we describe the information geometrical structure of mixture densities over totally bounded subsets of manifolds. This provides us with a computable parametric probability model over an arbitrarily large subset of Riemannian manifolds, and establishes a geometrical framework for population-based stochastic optimization and estimation over manifolds in the second part of the book.

The mixture densities are described by the simplex of mixture coefficients and mixture components consisting of locally inherited probability densities on the base manifold, both of which admit a statistical manifold structure. We then derive the information geometrical structure of the family of mixture densities, and show that it admits a product statistical manifold structure that is *separate* from the manifold structure of the base search space. This product statistical manifold of mixture densities thus provides the geometrical structure of the *decision space* over the Riemannian manifold search space.

- **[Part II: Model-Based Stochastic Derivative-Free Optimization on Riemannian Manifolds]**

  The product Riemannian structure of mixture densities over Riemannian manifolds provides us with a geometrical framework to tackle the optimization problem described in the beginning of Sect. 1.2. In the second part of the book we apply this framework to construct Extended RSDFO, a population-based stochastic derivative-free optimization algorithm on Riemannian manifolds.

  In Chap. 7, we survey the geometric aspects of two contemporary branches of optimization theories described in Sect. 1.1. We first review adaptations of optimization algorithms from Euclidean spaces to Riemannian manifolds, otherwise known as manifold optimization or Riemannian optimization in the literature. We

[6] Specifically, under the "geometrical" approach, we cannot pinpoint to a specific family of probability densities, whereas the "statistical approach is confined within a single normal neighbourhood.

then discuss the information geometric interpretation of population-based stochastic optimization algorithms on Euclidean spaces.

We show that Riemannian adaptations of Euclidean optimization algorithms to Riemannian manifolds all effectively employ the *same* principle as the "statistical" approach (of locally inherited probability densities over manifolds) described in Chap. 4. As a result the locality and assumptions of the "statistical" approach still persisted in manifold optimization algorithms, which in turn limits the generality of manifold optimization algorithms in the literature.

In order to overcome the locality and implicit assumptions of the Riemannian adaptation process, we require parametrized probability densities defined beyond the confines of a single normal neighbourhoods on Riemannian manifolds, and the mixture densities described in Chap. 6 provides exactly what is needed. This leads us to the discussion of the next chapter, where we overcome the local restrictions by proposing a population-based meta-algorithm using the geometrical framework described in Part I .

In Chap. 8, we first formalize a generalized framework, Riemannian Stochastic Derivative-Free Optimization (RSDFO) algorithms, for adapting Stochastic Derivative-Free Optimization (SDFO) algorithms from Euclidean spaces to Riemannian manifolds. RSDFO encompasses Riemannian adaptation of CMA-ES, and accentuates the main drawback of the Riemannian adaptation approach, i.e. the local restrictions. RSDFO also points us towards what is missing: parametrized probability densities defined beyond the confines of a single normal neighbourhood on Riemannian manifolds, whose statistical parameters are coherent with statistical estimations. This is accomplished by the notion of mixture densities over totally bounded subsets on Riemannian manifolds described in Chap. 6.

We then describe Extended RSDFO, a principled population-based meta-algorithm that uses existing manifold optimization algorithms, such as RSDFO, as its local module. Extended RSDFO addresses the local restriction of RSDFO using both the intrinsic Riemannian geometry of the manifold $M$ (search space) and the product statistical Riemannian geometry of families of mixture densities over $M$ (decision spaces). The components of Extended RSDFO are constructed from a geometrical perspective, and its properties derived rigorously from first principles. In particular, we describe the geometry of the evolutionary steps of Extended RSDFO using a "regularized inverse" Fisher metric on the simplex of mixture coefficients, and derive the convergence behaviors of Extended RSDFO on compact connected Riemannian manifolds.

In Chap. 9, we present and discuss several examples comparing Extended RSDFO with several state-of-the-art manifold optimization algorithms such as Riemannian Trust-Region method, Riemannian CMA-ES and Riemannian Particle Swarm Optimization on the $n$-sphere, Grassmannian manifold, and Jacob's ladder. Jacob's ladder, in particular, is a manifold of potentially infinite genus and cannot be addressed by traditional (constraint) optimization techniques on Euclidean spaces, which necessitates the development of manifold optimization algorithms.

Finally, we conclude the book and outline future research directions in Chap. 10.

## References


[AN00] S. Amari and H Nagaoka. *Methods of Information Geometry*, volume 191 of *Translations of Mathematical monographs*. Oxford University Press, 2000.

[Gab82] Daniel Gabay. Minimizing a differentiable function over a differential manifold. *Journal of Optimization Theory and Applications*, 37(2):177–219, 1982.

[SR+10] Gabriel Stoltz, Mathias Rousset, et al. *Free energy computations: A mathematical perspective*. World Scientific, 2010.

[CIDSZ08] Gunnar Carlsson, Tigran Ishkhanov, Vin De Silva, and Afra Zomorodian. On the local behavior of spaces of natural images. *International journal of computer vision*, 76(1):1–12, 2008.

[Whi44a] Hassler Whitney. The self-intersections of a smooth n-manifold in 2n-space. *Annals of Math*, 45(220-446):180, 1944.

[Whi44b] Hassler Whitney. The singularities of a smooth n-manifold in (2n- 1)-space. *Ann. of Math*, 45(2):247–293, 1944.

[KH05] John T Kent and Thomas Hamelryck. Using the fisher-bingham distribution in stochastic models for protein structure. *Quantitative Biology, Shape Analysis, and Wavelets*, 24:57–60, 2005.

[Oll93] Josep M Oller. On an intrinsic analysis of statistical estimation. In *Multivariate Analysis: Future Directions 2*, pages 421–437. Elsevier, 1993.

[Han06] Nikolaus Hansen. The cma evolution strategy: a comparing review. In *Towards a new evolutionary computation*, pages 75–102. Springer, 2006.

[FJLP03] P Thomas Fletcher, Sarang Joshi, Conglin Lu, and Stephen M Pizer. Gaussian distributions on lie groups and their application to statistical shape analysis. In *Biennial International Conference on Information Processing in Medical Imaging*, pages 450–462. Springer, 2003.

[AH19] P-A Absil and S Hosseini. A collection of nonsmooth riemannian optimization problems. In *Nonsmooth Optimization and Its Applications*, pages 1–15. Springer, 2019.

[ABG07] P-A Absil, Christopher G Baker, and Kyle A Gallivan. Trust-region methods on Riemannian manifolds. *Foundations of Computational Mathematics*, 7(3):303–330, 2007.

[AMS09] P-A Absil, Robert Mahony, and Rodolphe Sepulchre. *Optimization algorithms on matrix manifolds*. Princeton University Press, 2009.

[LL01] Pedro Larrañaga and Jose A Lozano. *Estimation of distribution algorithms: A new tool for evolutionary computation*, volume 2. Springer Science & Business Media, 2001.

[BA10] Pierre B Borckmans and Pierre-Antoine Absil. Oriented bounding box computation using particle swarm optimization. In *ESANN*, 2010.

[BIA10] Pierre B Borckmans, Mariya Ishteva, and Pierre-Antoine Absil. A modified particle swarm optimization algorithm for the best low multilinear rank approximation of higher-order tensors. In *International Conference on Swarm Intelligence*, pages 13–23. Springer, 2010.

[EK95] Russell Eberhart and James Kennedy. Particle swarm optimization. In *Proceedings of the IEEE international conference on neural networks*, volume 4, pages 1942–1948. Citeseer, 1995.

[CFFS10] Sebastian Colutto, Florian Fruhauf, Matthias Fuchs, and Otmar Scherzer. The cma-es on riemannian manifolds to reconstruct shapes in 3-d voxel images. *IEEE Transactions on Evolutionary Computation*, 14(2):227–245, 2010.

[HKK06] Thomas Hamelryck, John T Kent, and Anders Krogh. Sampling realistic protein conformations using local structural bias. *PLoS Computational Biology*, 2(9):e131, 2006.

[Pen04] Xavier Pennec. *Probabilities and statistics on Riemannian manifolds: A geometric approach*. PhD thesis, INRIA, 2004.

[ANOK12] Youhei Akimoto, Yuichi Nagata, Isao Ono, and Shigenobu Kobayashi. Theoretical foundation for CMA-ES from information geometry perspective. *Algorithmica*, 64(4):698–716, 2012.

# Part I
# Information Geometry of Probability Densities Over Riemannian Manifolds

# Chapter 2
# Riemannian Geometry: A Brief Overview

**Abstract** This chapter establishes the domain of discourse of the book and formally introduces the essential foundations of Differential Geometry with a flavour towards numerical computation. The chapter is divided into three parts: the first part introduces the notion of topological manifolds. The second part covers necessary objects such as vector bundles, Riemannian metric, and connections. The third part describes the main computational tools on Riemannian manifolds including exponential maps and normal neighbourhoods.

In this chapter we introduce the fundamentals of Riemannian geometry with a flavor towards numerical computation.

Modern differential geometry dates back to the works of Gauss and Riemann as an extension and generalization of classical Euclidean geometry, which studied objects such as curves and surfaces, and their "geometrical" properties such as curvature, angles and volume.

Differential geometry revolves around the notion of "manifolds", which generalizes "geometrical objects" such as curves, surfaces, spheres and Euclidean spaces studied in classical Euclidean geometry literature.

The central question of differential geometry is the identification, analysis and study of the *intrinsic* properties of geometrical objects. Intrinsic properties of an $n$-dimensional space $M$ are, in plain words [Lee06], are properties of $M$ that can be determined by a an $n$-dimensional entity living on $M$. Manifolds in this book are therefore considered to be "stand-alone" objects, free from an ambient Euclidean space. Equivalently and more formally, we study the intrinsic properties on manifolds that are invariant under isometry and reparametrization.

The material in this chapter summarizes the results discussed in the literature [Lee01, Lee06, Lor08, KN63, dC92, Lan12]. The goal of this chapter is to present the necessary objects and machineries in differential geometry, and to establish the foundation for both parts of the book. For detailed exposition of the relevant material, we refer to the aforementioned literature.

This chapter is divided into three sections. Section 2.1 establishes the domain of discourse: smooth topological manifolds. Section 2.2 revolves around topics of tangent spaces as a local linear approximation or a "linear scaffolding" of manifolds that

 
R. S. Fong and P. Tino, *Population-Based Optimization on Riemannian Manifolds*,
Studies in Computational Intelligence 1046,
https://doi.org/10.1007/978-3-031-04293-5_2

sets the framework of inheriting geometrical objects and properties from Euclidean spaces. Finally, we conclude with discussion of our main computational tools on Riemannian manifolds—exponential maps and normal neighbourhoods in Sect. 2.3.

## 2.1 Smooth Topological Manifolds

The most well-known and perhaps the simplest example of an $n$-dimensional geometrical object is the Euclidean spaces $\mathbb{R}^n$. Euclidean spaces are $n$-dimensional "flat" spaces that admits a global coordinate system. That is, all points admit a "vectorial" representation under the same system. Throughout this chapter Euclidean spaces will serve as a prototype to establish *local* geometrical objects and operations on manifolds.

We begin by describing the notion of manifolds: An $n$-dimensional topological manifold is a topological space that "looks like" a Euclidean space $\mathbb{R}^n$ locally. More formally:

**Definition 2.1** A topological space (with its corresponding topology) $M := (M, \mathcal{T}_M)$ is an $n$-**dimensional topological manifold** if $M$ satisfies:

1. **[Hausdorff]** $\forall x, y \in M, \exists U, V \in \mathcal{T}_M$ such that $x \in U$, $y \in V$ and $U \cap V = \phi$
2. **[Second countable]** $\mathcal{T}_M$ admits a countable basis, i.e. $M$ is not "too large".
3. **[Locally Euclidean of dimension** $n$**]** $\forall x \in M$ , $\exists U \in \mathcal{T}_M$ neighbourhood of $x$, and $V$ open subset of $\mathbb{R}^n$ such that $U$ is homeomorphic to $x$.

The local Euclidean property of (topological) manifolds means they can be locally viewed as a Euclidean space. The local linear identifications of manifolds with Euclidean spaces, formally known as local coordinate charts, can then be "stitched" together to obtain the global structure of the manifold.

**Definition 2.2** A **(smooth) local coordinate chart** is a pair $(U, \varphi)$ consisting of an open set $U \subseteq M$ and a (diffeomorphism) homeomorphism $\varphi : U \to V \subseteq \mathbb{R}^n$ onto an open subset $V$ of $\mathbb{R}^n$ called the **(smooth) local coordinate map**. The **local coordinates** on $U$ are the component functions $\left(x^1, \ldots x^n\right)$ of the local coordinate map $\varphi$ given by: for $x \in U$, $\varphi(x) = \left(x^1(x), \ldots, x^n(x)\right)$. Two local coordinate charts $(U, \varphi)$, $(V, \psi)$ are **smoothly compatible** if either

1. $U \cap V = \phi$ , OR
2. $U \cap V \neq \phi$ implies the **transition map**: $\psi \circ \varphi^{-1} : \phi(U \cap V) \to \psi(U \cap V)$ is a diffeomorphism on $\mathbb{R}^n$.

**Definition 2.3** A **smooth atlas** is a collection of pairwise smoothly compatible coordinate charts $\mathcal{E}_M = \{U_\alpha, \varphi_\alpha\}_{\alpha \in \Lambda_M}$ such that $U_\alpha$'s cover $M$. In other words, $M = \bigcup_{\alpha \in \Lambda_M} U_\alpha$. An atlas $\mathcal{E}_M$ is **maximal** if it is not contained in any larger smooth atlas.

**Definition 2.4** A **smooth manifold** is a pair $(M, \mathcal{E}_M)$ consisting of a topological manifold $M$ and a maximal smooth atlas $\mathcal{E}_M$.

## 2.2 Tangent Spaces, Metric and Curvature

The local Euclidean property of topological manifolds allows us to inherit topology from Euclidean spaces locally via coordinate charts. However, though invariant under diffeomorphism, the local coordinate charts only provide us with *one* (local) representation of the manifold with a particular choice of local coordinate system. In order to discuss the *intrinsic* geometrical properties of manifolds, we start from the manifold and consider its tangent spaces—local linear approximation of the manifold.

Just as smooth curves in $\mathbb{R}^n$ admit local linear approximations by tangent vectors, manifolds admit a form of linear approximation in the form of tangent space. We begin by describing tangent spaces of $\mathbb{R}^n$.

Given $x \in \mathbb{R}^n$, the tangent space centered at $x$ consists of all vectors originating at $x$:

$$T_x\mathbb{R}^n := \left\{x + v \middle| v \in \mathbb{R}^n\right\} \cong \mathbb{R}^n.$$

Consider a curve $\gamma : \mathbb{R} \to \mathbb{R}^n$ through $x$ in the direction of tangent vector $v \in T_x\mathbb{R}^n$: $\gamma(t) = x + tv$, where $t \in \mathbb{R}$ denote the standard coordinates. Let $f : \mathbb{R}^n \to \mathbb{R}$ be a smooth function defined on a neighborhood $U$ of $x \in \mathbb{R}^n$, then the **directional derivative** of $f$ at direction $v$ at $x$ is given by:

$$D_v|_x f = \left.\frac{d}{dt}\right|_{t=0} f(\gamma(t)) = \left.\frac{d}{dt}\right|_{t=0} f(x + tv).$$

In coordinates $\left(x^1, \ldots, x^n\right)$ of $\mathbb{R}^n$, we may write $v = \left(v^1, \ldots, v^n\right)$, and the above equation becomes:

$$D_v|_x f = \sum_{i=1}^{n} v^i \frac{df}{dx^i}(p).$$

Let $C^\infty(\mathbb{R}^n)$ denote the set of real-valued smooth functions on $\mathbb{R}^n$, the equation above defines a linear operator for each $x \in \mathbb{R}^n$, denoted by $D_v|_x$:

$$D_v|_x = \sum_{i=1}^{n} v^i \frac{d}{dx^i}. \tag{2.1}$$

If two (differentiable) functions agree on a small neighbourhood of $x$ then their directional derivatives at $x$ would agree. More formally:

**Definition 2.5** For each point $x \in \mathbb{R}^n$, a **smooth function elements** through $x \in \mathbb{R}^n$ is a pair $(f, U)$, where $U \subset M$ is open subset containing $x$ and $f : U \to \mathbb{R}$ is smooth. The set of smooth function elements through $x$ on $\mathbb{R}^n$ is denoted by $\overline{C}_x^\infty(\mathbb{R}^n)$.

Furthermore, we say that $(f, U)$ and $(g, V)$ are equivalent if and only if they agree on some neighbourhood containing $x$:

$$(f, U) \sim (g, V) \leftrightarrow \exists W \subset U \cap V \text{ s.t. } f|_W \equiv g|_W .$$

Given a smooth function element $(f, U)$ through $x \in \mathbb{R}^n$, the **germ of** $f$ **at** $x$ is the equivalent class in the quotient space: $C_x^\infty(\mathbb{R}^n) := \overline{C}_x^\infty(\mathbb{R}^n)\backslash \sim$.

Using the above definition, it is clear that the directional derivative operator $D_v|_x$ in Eq. (2.1) is defined on the equivalent class of germs of (smooth) functions through $x$. That is, for each $x \in \mathbb{R}^n$, $D_v|_x : C_x^\infty(\mathbb{R}^n) \to \mathbb{R}$, and more generally:

**Definition 2.6** A **derivation** at $x$ on $\mathbb{R}^n$ is a linear map $X : C_x^\infty(\mathbb{R}^n) \to \mathbb{R}$ that satisfies the Leibniz rule (product rule):

$$X(fg) = f \cdot X(g) + g \cdot X(f) \tag{2.2}$$

where $f, g \in C_x^\infty(\mathbb{R}^n)$ are germs at $x$.

There's a natural identification of tangent vectors with directional derivatives:

**Theorem 2.1** *The set of derivatives through $x \in \mathbb{R}^n$ is a vector space isomorphic to the tangent space $T_x\mathbb{R}^n$ at $x$ via the map:*

$$v \in T_x\mathbb{R}^n \mapsto D_v|_x = \sum_{i=1}^n v^i \frac{d}{dx^i}$$

**Remark 2.1** The importance of the above characterization is twofolds:

1. It demonstrates the local nature of tangent spaces. Tangent vectors are constructed on local equivalent classes of functions through a point. This characteristic carries onto the manifold case as well.
2. The geometrically intuitive tangent vectors can be realized more abstractly through the notion of derivations. More importantly, the notion of derivations is *intrinsic*, in the sense that they are invariant under reparametrizations of local coordinates.

The same arguments can be repeated to construct tangent vectors on manifolds.

**Definition 2.7** Let $M$ be an $n$-dimensional manifold, a function $f : M \to \mathbb{R}^k$ is **smooth** at $x \in M$ if there is a local coordinate chart $(U, \varphi)$ containing $x$, such that $f \circ \varphi^{-1} : \mathbb{R}^n \to \mathbb{R}^k$ is smooth. The set of smooth real-valued functions $f : M \to \mathbb{R}$ is denoted by $C^\infty(M)$.

**Definition 2.8** A **derivation** or **tangent vector** at $x \in M$ is a linear map $X : C_x^\infty(M) \to \mathbb{R}$ that satisfies the Leibniz rule (Eq. (2.2)).

The **tangent space** of $M$ at $x$, denoted by $T_x M$ is the set of all derivations on $M$ at $x$.

### 2.2.1 Tangent Bundle and Riemannian Metric

Tangent spaces of an $n$-dimensional manifold are $n$-dimensional vector spaces. Additional structures can be equipped on each tangent space to describe geometrical properties such as angle and distance locally. These local information can then be be "stitched" together to form globalized counterparts through out the entire manifold. It is therefore natural to study tangent spaces collectively, rather than as disjoint vector spaces attached to points on the manifold. The collection of tangent spaces is known as the tangent bundle of $M$.

We begin by describing vector bundles, a slightly generalized notion of tangent bundles, where we attach to each point $x \in M$ a local vector space instead of tangent space. This generalization is necessary to describe the geometrical objects introduced later into the chapter.

**Definition 2.9** A **(smooth) vector bundle** of rank $k$ is a pair of (smooth) manifolds $(E, M)$ along with a (smooth) continuous surjective map $\pi : E \to M$.

$(E, M, \pi)$[1] are **total space**, **base**, and **projection** respectively satisfying:

1. $E_x := \pi^{-1}(x)$ (**fibre of** $E$ **at** $x$) is a vector space.
2. $\forall x \in M, \exists U$ a neighbourhood of $x$, such that the following diagram commutes with homeomorphism (diffeomorphism for smooth bundles) (**local trivialization**) $\varphi : \pi^{-1}(U) \to U \times \mathbb{R}^k$

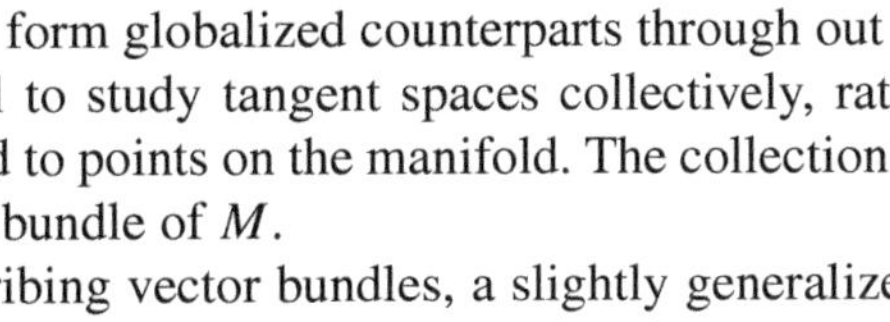

$$\begin{array}{ccc} \pi^{-1}(U) \subset E & \xrightarrow{\quad\varphi\quad} & U \times \mathbb{R}^k \\ {\scriptstyle\pi}\searrow \;\nwarrow{\scriptstyle\sigma} & & \swarrow{\scriptstyle\pi_1} \\ & U \subset M & \end{array}$$

3. $\varphi_x : \pi^{-1}(x) = E_x \to \{x\} \times \mathbb{R}^k$ is a linear isomorphism.

**Definition 2.10** A **(smooth) section** of $E$ is a (smooth) continuous map $\sigma : M \to E$ such that $\pi \circ \sigma = Id_M$. Equivalently $\sigma(x) \in E_x$ for all $x$. The space of smooth sections of $E$ is denoted by $\mathcal{E}(M)$.

In the special case when the local vector spaces are tangent spaces, we have the notion of tangent bundle:

**Definition 2.11** **Tangent bundle** of $M$ is a vector bundle $(TM, M)$ of rank $n$ defined by:

$$TM = \bigsqcup_{x \in M} T_x M = \bigcup_{x \in M} \{x\} \times T_x M$$

The space of smooth sections of $TM$, denoted by either $\mathcal{E}(TM)$ or $\mathcal{T}M$ is the space of smooth **vector fields** on $M$ that maps each point $x \in M$ to a tangent vector $V_x \in T_x M$:

[1] A vector bundle is sometimes referred to as $\pi : E \to M$.

$$
\begin{aligned}
V : M &\to TM \\
x &\mapsto V_x \in T_x M
\end{aligned}
$$

Given two manifolds, we can translate tangent vectors from one to another.

**Definition 2.12** Consider manifolds $M$, $N$ and their corresponding tangent bundles $\pi_M : TM \to M, \pi_N : TN \to N$. Given smooth map $f : M \to N$, the **pushforward** of $f$, denoted by $f_*$ is defined as follows: For each $x \in M$, the map $f_* : T_x M \to T_{f(x)} N$ is given by:

$$
(f_* X)(g)|_{f(x)} = X(g \circ f)|_x ,
$$

where $X \in T_x M$, $g \in C^\infty(N)$, and $f_* X \in T_{f(x)} N$. In other words, the bundle morphism under the following diagram commutes:

$$
\begin{array}{ccc}
TM & \xrightarrow{f_*} & TN \\
\pi_M \downarrow & & \downarrow \pi_N \\
M & \xrightarrow{f} & N
\end{array}
$$

Let $(U, \varphi)$ be a smooth local coordinate chart of $M$ containing $x \in M$. Let $(x^1, \ldots, x^n)$ denote the local coordinates corresponding to the chart. We can derive a basis of $T_x M$ corresponding to the local coordinates by pushing forward the basis from $\mathbb{R}^n$.

The pushforward of the local coordinate map is given by: $\varphi_* : T_x M \to T_{\varphi(p)} \mathbb{R}^n$. Consider basis of $T_{\varphi(x)} \mathbb{R}^n$ given by the set of derivations $\left( \frac{\partial}{\partial x^n}\big|_{\varphi(x)}, \ldots, \frac{\partial}{\partial x^n}\big|_{\varphi(x)} \right)$. Since $\varphi$ is a diffeomorphism, the pushforward $\varphi_*$ is a linear isomorphism thus invertible. Therefore we obtain a basis of $T_x M$ by:

$$
\left( \frac{\partial}{\partial x^1}\bigg|_x, \ldots, \frac{\partial}{\partial x^n}\bigg|_x \right) := \left( \left(\varphi^{-1}\right)_* \frac{\partial}{\partial x^n}\bigg|_{\varphi(x)}, \ldots, \left(\varphi^{-1}\right)_* \frac{\partial}{\partial x^n}\bigg|_{\varphi(x)} \right)
$$

This basis of $T_x M$ is call the **local coordinate frame on** $U$ corresponding to the coordinate system. For any $x \in M$, a tangent vector $v \in T_x M$ can be written uniquely as the linear combination:

$$
v = \sum_{i=1}^{n} v^i \frac{\partial}{\partial x^i}\bigg|_x ,
$$

where $v^i \in \mathbb{R}$ for $i = 1, \ldots, n$. Furthermore, this can be extended to form a set of $n$ linearly independent smooth sections on $TM$: For $i = 1, \ldots, n$:

$$
\begin{aligned}
\partial_i := \frac{\partial}{\partial x^i} : M &\to TM \\
x &\mapsto \frac{\partial}{\partial x^i}\bigg|_x .
\end{aligned}
$$

The set $(\partial_i)_{i=1}^n := \left(\frac{\partial}{\partial x^i}\right)_{i=1}^n$ is called a **coordinate frame**. With a slight abuse of notation, a smooth vector field $v$ (smooth section of $TM$) can thus be expressed as:

$$v = \sum_{i=1}^{n} v^i \partial_i,$$

where $v^i : M \to \mathbb{R}$ are smooth real-valued functions for each $i = 1, \dots, n$.

#### 2.2.1.1 Riemannian Metric

Under the notion of tangent bundle, we can "mount" additional structure on the manifold through the tangent spaces. One of the most important structures is the notion of metric:

**Definition 2.13** A **Riemannian metric** $g$ on $M$ is a **symmetric**, **positive definite**, **bilinear** map defined on each $T_xM$ (more formally a $\binom{2}{0}$ tensor field):

$$\begin{aligned} g : T_xM \times T_xM &\to M \\ (X, Y) &\mapsto g(X, Y) =: \langle X, Y \rangle_g. \end{aligned}$$

**Definition 2.14** A **Riemannian manifold** is a pair $(M, g)$, where $M$ is a manifold and $g$ is a Riemannian metric on $M$.

Riemannian metric is an extremely powerful tool on manifolds. In particular, it allows us to define angles and norm of tangent vectors. The norm of tangent vectors in turn provides us with the notion of length of curves on manifolds (via the tangent of the curve). On connected Riemannian manifolds, this gives us a Riemannian distance[2] between *any* two points on the manifold. More importantly, the induced metric space topology by the Riemannian distance coincides with the manifold topology! The detailed exposition is beyond the scope of the book and we refer to the aforementioned literature.

### 2.2.2 *Affine Connection and Parallel Transport*

At this point of the discussion, the tangent spaces in the tangent bundle remains disjoint and the geometrical properties of manifolds are inherited only **locally**. In order to relate the local geometric structures the set of tangent spaces is discussed under one single set—the tangent bundle, where each fibre is homeomorphic to the same topological space (in $TM$ they are all linearly isomorphic to $\mathbb{R}^n$). For the rest of

[2] It is worth noting that whilst this distance exists theoretically, it maybe computationally infeasible.

the discussion on tangent spaces, we discuss how we can move between the disjoint tangent spaces of the tangent bundle.

We once again draw inspiration from the Euclidean space $\mathbb{R}^n$, which has two distinctive properties:

1. For any $x \in \mathbb{R}^n$, $T_x\mathbb{R}^n \cong \mathbb{R}^n$. There is a *global* isomorphism between the tangent space and the base space.
2. The global coordinate system of $\mathbb{R}^n$ induces a *global coordinate frame*. Every tangent vector at any point of $\mathbb{R}^n$ can be expressed as linear combinations of the basis of *one* tangent space (for example the standard basis $\{e_i\}$ of $T_0\mathbb{R}^n$.)

Given tangent vectors $X, Y \in T_x\mathbb{R}^n$ at $x$, we can "project" $Y$ along the direction of $X$ via directional derivative:

$$\nabla_X Y|_x = \lim_{t\to 0} \frac{Y_{x+t\cdot v_x} - Y_x}{t}. \tag{2.3}$$

On abstract manifold $M$ we run into two problems corresponding to the two properties of $\mathbb{R}^n$ discussed in the previous paragraph respectively:

1. There is no global diffeomorphism between the tangent spaces of $M$ and the manifold $M$.[3] Therefore the term $Y_{x+t\cdot v_x}$, specifically the subscript $x + t \cdot v_x$ is not well defined.
2. Tangent spaces of $M$ are disjoint spaces and there is no "natural" way to take the quotient $Y_{x+t\cdot v_x} - Y_x$.

Nevertheless this gives us an idea to "relate local geometry information" by finding a way to map one tangent space to another via a notion analogous to covariant derivative in the context of vector fields (smooth sections). Hence the notion and the name of **connections**.

**Definition 2.15** Let $(E, M, \pi)$ be a vector bundle over manifold $M$, a **connection** in $E$ is the map:

$$\begin{aligned} \nabla : \mathcal{T}M \times \mathcal{E}(M) &\to \mathcal{E}(M) \\ (X, Y) &\mapsto \nabla_X Y, \end{aligned}$$

where $\mathcal{E}(M)$ denote the smooth sections of $E$ and $\mathcal{T}M$ are smooth sections of the tangent bundle. The map $\nabla$ satisfies:

1. $C^\infty(M)$**-linear in** $X$

$$\nabla_{f\cdot X_1 + g\cdot X_2} Y = f \cdot \nabla_{X_1} Y + g \cdot \nabla_{X_2} Y, \quad \forall f, g \in C^\infty(M), \forall X_1, X_2 \in \mathcal{T}M$$

---

[3] However, a local diffeomorphism does exist, we shall elaborate this further in the the discussion of exponential maps.

2. **Linear over $\mathbb{R}$ in $Y$**

$$\nabla_X (aY_1 + bY_2) = a \cdot \nabla_X Y_1 + b \cdot \nabla_X Y_2, \quad \forall a, b \in \mathbb{R}, \forall Y_1, Y_2 \in \mathcal{E}(M)$$

3. **Leibniz (product) rule:**

$$\nabla_X(f \cdot Y) = Xf \cdot Y + f \cdot \nabla_X Y, \quad \forall f \in C^\infty(M)$$

$\nabla_X Y$ is called the **covariant derivative of $Y$ in direction $X$**.

By definition, given $x \in M$, $\nabla_X Y$ depends only on $Y$ on some neighbourhood of $x$, and $X$ at the point $x$. When restricting our attention to the tangent bundle $TM$ over $M$ as the vector bundle, we obtain.

**Definition 2.16** An **affine connection** on $M$ is the connection in $TM$:

$$\nabla : \mathcal{T}M \times \mathcal{T}M \to \mathcal{T}M,$$

where $\mathcal{T}M$ are smooth sections of the tangent bundle, i.e. smooth vector fields on $M$.

Let $U$ be an open subset of $M$, suppose $\{E_i\}_{i=1}^n$ is a local frame (linearly independent sections) of $TM$ on $U$. For each pair of indices $i$, $j$, we can express $\nabla_{E_i} E_j$ as:

$$\nabla_{E_i} E_j = \sum_{i,j=1}^{n} \Gamma_{i,j}^k E_k,$$

where $\Gamma_{i,j}^k$ is a set of $n^3$ functions called the **Christoffel symbols (of the second kind)**. An affine connection can be completely described by a set of Christoffel symbols: Given $U \subset M$, let $\{E_i\}_{i=1}^n$ be a local frame of $TU \subset TM$. Vector fields $X, Y \in \mathcal{T}U$ can be expressed as $\sum_i X^i E_i$, $\sum_j Y^j E_j$ respectively, and we have:

$$\nabla_X Y = \sum_{i,j,k=1}^{n} \left( X^i E_i Y^k + X^i Y^j \Gamma_{i,j}^k \right) E_k. \tag{2.4}$$

On Euclidean space, we retrieve the directional derivative on $M = \mathbb{R}^n$ via the **Euclidean connection** given by:

$$\overline{\nabla}_X Y = \sum_{j=1}^{n} XY^j E_j.$$

In other words, the Christoffel symbols of the Euclidean connection in $\mathbb{R}^n$ vanish identically in standard coordinates.

### 2.2.3 Parallel Transport

From the discussion above, affine connections provide us with a theoretical way to move from one tangent space to another. In this section we realize the transition with the notion of **parallel transport**. This will allow us to retrieve the directional derivative of Eq. (2.3) by the end of the section.

**Definition 2.17** Let $M$ be a manifold. A vector field $V \in \mathcal{T}M$ is **parallel** if $\nabla_X V \equiv 0$ for all $X \in \mathcal{T}M$.

In general, non-zero parallel vector fields do not necessarily exist over the entire manifold $M$. On the other hand, we may construct parallel vector fields along a curve[4] in $M$:

**Definition 2.18** Given a curve $\gamma : I \subset \mathbb{R} \to M$, a **(smooth) vector field along** $\gamma$ is a (smooth) map $V : I \to TM$ such that $V(t) \in T_{\gamma(t)}M$. A vector field $V$ along $\gamma : I \to M$ is **parallel along** $\gamma$ if $\nabla_{\dot{\gamma}(t)} V \equiv 0$ for all $t \in I$, where $\dot{\gamma}(t) := \sum_i \frac{d}{dt}\gamma^i(t)E_i$ for some local frame $\{E_i\}$.

**Theorem 2.2** *Given a curve $\gamma : I \subset \mathbb{R} \to M$ and a tangent vector $V_0 \in T_{\gamma(t_0)}M$ at $t_0 \in I$, there exists a unique parallel vector field $V$ along $\gamma$, such that $V(t_0) = V_0$.*

$V$ is called **parallel translation of $V_0$ along** $\gamma$, and it defines an important operator: a natural linear isomorphism between tangent spaces.

**Definition 2.19** Given a curve $\gamma : I \to M$ and $t_0, t_1 \in I$. **Parallel transport** from $T_{\gamma(t_0)}M$ to $T_{\gamma(t_1)}M$ is the linear isomorphism:

$$P_{t_0,t_1} : T_{\gamma(t_0)}M \to T_{\gamma(t_1)}M,$$

such that for any $t_1 \in I$ $P_{t_0,t_1}$ satisfies:

$$P_{t_0,t_1} V_0 = P_{t_0,t_1} V(t_0) = V(t_1),$$

where $V_0 \in T_{\gamma(t_0)}M$ and $V$ denote the parallel translation of $V_0$ along $\gamma$.

Finally, we retrieve a formula of covariant derivatives in $M$ very much similar to Eq. (2.3) on $\mathbb{R}^n$:

**Lemma 2.1** *Let $V \in \mathcal{T}(\gamma)$ be a vector field along $\gamma$. The covariant derivative $\nabla_{\dot{\gamma}(t)} V(t)$ along $\gamma$ can be expressed as:*

$$\nabla_{\dot{\gamma}(t)} V(t)\Big|_{t=t_0} = \lim_{t \to t_0} \frac{P_{t_0,t}^{-1} V(t) - V(t_0)}{t - t_0} \tag{2.5}$$

[4] Without loss of generality, we may assume the curves are injective.

***Proof*** Let $V \in \mathcal{T}(\gamma)$ be a vector field along $\gamma$. Let $\{x_1, \ldots, x_n\}$ denote the set of local coordinates in a neighbourhood of $\gamma(t_0)$, then we can write:

$$V(t) = \sum_{j=1}^{n} V^j(t)\partial_j.$$

By Theorem 2.2, we extend $\{\partial_j\}$ to a a parallel frame (of vector fields) $\{\tilde{\partial}_j(t)\}$ along $\gamma$, this implies $\nabla_{\gamma(t)}\tilde{\partial}_j \equiv 0$. We then have the following expansion (by Eq. (2.4)):

$$\begin{aligned}
\nabla_{\dot{\gamma}(t)}V(t)\big|_{t=t_0} &= \sum_{j=1}^{n} \dot{V}^j(t_0)\partial_j + V^j(t_0)\underbrace{\nabla_{\gamma(t_0)}\partial_j}_{=0} = \sum_{j=1}^{n} \dot{V}^j(t_0)\partial_j \\
&= \sum_{j=1}^{n} \lim_{t\to t_0} \frac{V^j(t) - V^j(t_0)}{t - t_0}\partial_j = \sum_{j=1}^{n} \lim_{t\to t_0} \frac{V^j(t)\partial_j - V^j(t_0)\partial_j}{t - t_0} \\
&= \sum_{j=1}^{n} \lim_{t\to t_0} \frac{P_{t_0,t}^{-1}V^j(t)\tilde{\partial}_j(t) - V^j(t_0)\partial_j}{t - t_0} = \lim_{t\to t_0} \frac{P_{t_0,t}^{-1}V(t) - V(t_0)}{t - t_0}.
\end{aligned}$$

□

## 2.3 Domain of Computation: Exponential Map and Normal Neighbourhood

In this section we discuss the "domain of computation" of the book on Riemannian manifolds. Computation on the entire abstract Riemannian manifolds is generally difficult, this is primarily due to the lack of global parametrization and the elusiveness of global properties.

From the construction of smooth manifolds, the local structure is implicitly induced from Euclidean spaces via local coordinate charts. One might therefore attempt to perform computation on the Euclidean spaces first, and then map the results back to the manifold via coordinate charts. However, important geometrical properties such as distance and convexity are not preserved under the local coordinate maps as they are "just" diffeomorphisms. Moreover, coordinate charts covering the manifold are disjoint spaces and can only be related in their intersections.

On the other hand, tangent space provides us with an intrinsic linear approximation of the manifold. Given two points connected by a parametric curve, we discussed how tangent vectors within two disjoint tangent spaces can be translated from one to another via parallel transport. Moreover, geometrical properties such as angles and curvature can be studied locally through the tangent spaces.

On Euclidean spaces, for any $x \in \mathbb{R}^n$, there is a natural identification of the base space and its tangent spaces. In particular, $\mathbb{R}^n \cong T_x\mathbb{R}^n$ for any $x$. On Riemannian manifolds, this relation is realized locally through **Riemannian exponential map**.

Riemannian exponential map is an extremely important computational tool on Riemannian manifolds. It allows computations on Riemannian manifolds to be performed locally on tangent spaces in the Euclidean fashion. The results are subsequently mapped back to the manifold via the local Riemannian exponential map preserving the local geometrical information.

We begin the discussion by describing "straight lines" on Riemannian manifolds. In the Euclidean case, a curve $\gamma : I \subset \mathbb{R} \to \mathbb{R}^n$ is a straight line if it has zero acceleration. In other words, the parallel translation of the velocity of $\gamma$ along itself remains unchanged. Formally, let $\{t\}$ denote the standard coordinate of $I \subset \mathbb{R}$, the velocity of $\gamma$ in $M$ is given by: $\dot{\gamma} := \gamma_* \frac{d}{dt}$ (pushforward of the tangent vector $\frac{d}{dt}$ of $I \subset \mathbb{R}$). A curve $\gamma$ in $\mathbb{R}^n$ has zero acceleration if: $\nabla_{\dot{\gamma}}\dot{\gamma} \equiv 0$.

The above discussion is summarized by the following definition of geodesic curves on Riemannian manifold $M$:

**Definition 2.20** Let $(M, g)$ be a Riemannian manifold with affine connection $\nabla$. A curve $\gamma : I \subset \mathbb{R} \to M$ is called a **geodesic** with respect to $\nabla$ if it has zero acceleration:

$$\nabla_{\dot{\gamma}}\dot{\gamma} \equiv 0.$$

For sufficiently small $I_v$, the geodesic is the curve on $(M, g)$ that connects $x$ and $\gamma_v(I_v)$ with a path of minimum length with respect to the Riemannian distance generated by Riemannian metric $g$.

Moreover, given a point $x$ in Riemannian manifold $(M, g)$ and a tangent vector $v \in T_xM$, there exists a unique geodesic $\gamma_v : [0, I_v) \subset \mathbb{R} \to M$ with initial point $x$ and initial velocity $v \in T_xM$. In other words, given a starting point $x \in M$ and an initial velocity, the geodesic with $\gamma_v(0) = x$ and $\dot{\gamma}_v(0) = v$ is uniquely determined.

This uniqueness of geodesic provides us with a way to map tangent vectors centered at a point $x \in M$ locally to the base manifold. In particular, given a point $x \in M$ and $v \in T_xM$, we have a unique point in $M$ by tracing along $\gamma_v$ starting at $x$ with initial velocity $v \in T_xM$ for time 1. Formally, we have the following definition:

**Definition 2.21** Given a point $x \in M$, consider the subset $O_x$ of $T_xM$ given by:

$$O_x := \left\{ v \in T_xM \middle| \gamma_v \text{ is defined on } [0, I_v), I_v > 1 \right\}.$$

The **Riemannian exponential map** at $x$ is the map:

$$\begin{aligned} \exp_x : O_x \subset T_xM &\to M \\ \exp_x(v) &:= \gamma_v(1). \end{aligned}$$

For each $x \in M$, there exists a open star-shaped neighbourhood $U_x$ around $\mathbf{0} \in T_xM$ where the Riemannian exponential map is a *local* diffeomorphism. The image

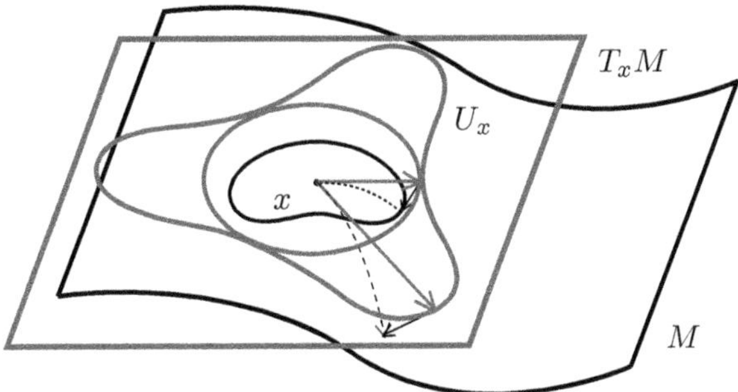


**Fig. 2.1** Illustration of normal neighbourhood and geodesic ball on the manifold $M$. The green "flat" space attached to the point $x \in M$ represents $T_x M$, while the star-shaped neighbourhood represents $U_x = \exp_x^{-1}(N_x)$. The figure shows tangent vectors within the green disc (ball of injectivity radius) gets mapped to a the geodesic ball (black loop) on $M$

of $U_x$ under $\exp_x$ is the open neighbourhood $N_x := \exp_x(U_x) \subset M$ of $x \in M$ called the **normal neighbourhood** of $x$.

Moreover, since $\exp_x : U_x \to N_x$ is a local diffeomorphism, for each $y \in \exp_x(U_x)$ there is a unique $v \in U_x$ such that $\exp_x(v) = y$. In particular, the Riemannian exponential map is locally invertible and the inverse is called the **Riemannian logarithm map**:

$$\log_x := \exp_x^{-1} : N_x \to U_x.$$

It is important to note that Riemannian logarithm map *only* exists locally within the normal neighbourhood. The discussion of this section is summarized in Fig. 2.1.

### *2.3.1 Parallel Transport in Geodesic Balls/Normal Neighborhoods*

Let $M$ be a Riemannian manifold. For any point $x \in M$, the normal neighborhood $N_x$ of $x$ is diffeomorphic to a star-shaped open neighbourhood centered at $\mathbf{0} \in T_x M$ via the Riemannian exponential map. Local computations on Riemannian manifolds can therefore be carried out on the tangent space in a similar fashion as in the Euclidean case, and the results subsequently translated back onto the manifold locally via the Riemannian exponential map.

The algorithms discussed in this book will therefore focus on normal neighbourhoods centered around the search iterates in the manifold $M$. In particular, we will use parallel transport to transfer search information from (the tangent space of) the current iterate to the next, within the normal neighbourhood of the current iterate.

Given a point $x \in M$ and a normal neighbourood $N_x$ of $x$, let $\{e_1(x), \ldots, e_(x)\}$ denote an orthonormal basis for $T_x M$[5]. This induces a linear isomorphism: $E : \mathbb{R}^n \to$

[5] This can be generated using the Gram–Schmidt process.

$T_xM$ mapping $(v^1, \ldots, v^n) \mapsto \sum_{i=1}^{n} v^i e_i(x)$. Together with the Riemannian exponential map, we obtain a coordinate function within $N_x$, called **normal coordinates centered at** $x$, given by: $E^{-1} \circ \exp_x^{-1} : N_x \to \mathbb{R}^n$.

Since $\exp_x$ is a diffeomorphism with range $N_x$, any point $y$ in the normal neighbourhood $N_x$ centered at $x$ can be connected to $x$ via a unique geodesic starting at $x$ with some initial velocity $v$. More formally, for any $y \in N_x$ we can choose $v \in T_xM$ such that $y = \exp_x(v)$. Within the normal coordinate centered at $x$, this geodesic is represented by a radial line segment radiating from $x$ in $N_x$: let $v = \sum_{i=1}^{n} v^i e_i(x) \in T_xM$, the geodesic $\gamma_v(t) : [0, 1) \to M$ with initial velocity $v$ starting at $x$ is given by the radial line segment [Lee06]:

$$\gamma_v(t) = \left(tv^1, \ldots, tv^n\right). \tag{2.6}$$

The tangent space of any point in $N_x$ thus admits an orthonormal basis corresponding to the $N_x$ and its Riemannian exponential map. Given $x, y \in N_x \subseteq M$, let $P_{x,y} : T_xM \to T_yM$ denote the parallel transport from $T_xM$ to $T_yM$ along $\gamma_v$, and $\{e_1(x), \ldots, e_n(x)\}$ denote an orthonormal basis for $T_xM$. Since parallel transport is an isometry, the set of tangent vectors $\{e_1(y), \ldots, e_n(y)\} := \left\{P_{x,y}e_1(x), \ldots, P_{x,y}e_n(y)\right\}$ is an orthonormal basis for $T_yM$. Together with the radial geodesic described in Eq. (2.6), the parallel transport of any tangent vector $w := \sum_{i=1}^{n} w^i e_i(x) \in T_xM$ from $T_xM$ to $T_yM$, *within the normal neighbourhood* $N_x$, is given by:

$$P_{x,y}w = \sum_{i=1}^{n} w^i e_i(y) \in T_yM \tag{2.7}$$

The above discussion is summarized in Fig. 2.2.

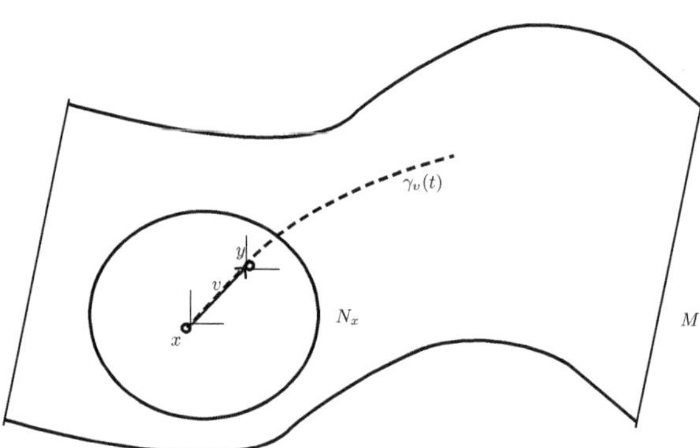


**Fig. 2.2** Illustration of parallel transport from $x$ to $y$ within normal neighbourhood $N_x$ of $x$, described in Sect. 2.3.1. The orthonormal basis at $x$ is parallel transported to an orthonormal basis at $y$

Finally, we consider metric balls in the pre-image of the normal neighbourhood, which are mapped to metric balls in $M$ under normal coordinates.

**Definition 2.22** Let $M$ be a Riemannian manifold. Given a point $x \in M$, the **injectivity radius** at $x$ is the real number [Pet06]:

$$\mathrm{inj}(x) := \sup_{r \in \mathbb{R}} \left\{ \exp_x : B(\mathbf{0}, r) \subset T_x M \to M \text{ is a diffeomorphism} \right\},$$

where $B(\mathbf{0}, r)$ is a ball of radius $r$ centered at $\mathbf{0}$ in $T_x M$.

For any $x \in M$, the **ball of injectivity radius**, denoted by $B(\mathbf{0}, \mathrm{inj}(p)) \subset U_x \subset T_x M$, is the largest metric ball centered at $\mathbf{0} \in T_x M$ such that the Riemannian exponential map $\exp_x$ is a diffeomorphism.

For any $j_x \leq \mathrm{inj}(x)$, the set $\exp_x (B(\mathbf{0}, j_x)) \subset M$ is a neighbourhood of $x \in M$ called a **geodesic ball**.

**Remark 2.2** Geodesic balls in $M$ are also metric balls in $M$ of the same radius. For the rest of the book, the geodesic balls are *closed* unless specified otherwise.

## 2.4 Discussion and Outlook

This chapter overviews the essential objects in modern differential geometry, both theoretical and computational, discussed in the literature [Lee01, Lee06, Lor08, KN63, dC92, Lan12]. We begin our discussion by surveying the notion of smooth topological manifolds, vector bundles and affine connections, which establishes the necessary vocabulary for the discussions in both parts of the book. This lays the foundation for the geometry of both the *decision space* and the *search space* of the manifold optimization problem described in Sect. 1.2.

In the next chapter, we will look closer into the geometrical structure of the *decision space* of the manifold optimization problem (Sect. 1.2) which consists of finitely parametrized probability densities over Riemannian manifolds. In particular, we discuss aspects of the emerging field of Information Geometry—the study of differential geometrical structures of finitely parametrized statistical models.

## References


[dC92] M.P. do Carmo. *Riemannian Geometry*. Mathematics (Boston, Mass.). Birkhäuser, 1992.

[Lee01] John M Lee. *Introduction to smooth manifolds*. Springer, 2001.

[Lee06] John M Lee. *Riemannian manifolds: an introduction to curvature*, volume 176. Springer Science & Business Media, 2006.

[Pet06] Peter Petersen. *Riemannian geometry*, volume 171. Springer, 2006.
[Lan12] Serge Lang. *Fundamentals of differential geometry*, volume 191. Springer Science & Business Media, 2012.
[KN63] Shoshichi Kobayashi and Katsumi Nomizu. *Foundations of differential geometry*, volume 1. New York, 1963.
[Lor08] W Tu Loring. An introduction to manifolds, 2008.

# Chapter 3
# Elements of Information Geometry

**Abstract** This chapter describes the geometrical structure of statistical models studied in the emerging field of Information Geometry, formally known as statistical manifolds. Due to the dual nature of statistical manifolds as both a statistical model and a Riemannian manifold, it requires a specialized geometrical structure that is different from the ones studied in "classical" Differential Geometry. In this chapter, we describe the notion of statistical manifolds and the relevant objects such as Fisher-Rao information metric, dualistic structure, and divergence. The chapter also discusses notions of Levi-Civita connection and Riemannian curvature from classical Differential Geometry, and describe their extensions to the context of statistical manifolds.

In this chapter we discuss the emerging field of Information Geometry—the study of statistical manifolds, the Riemannian manifolds of families of finitely parametrized probability densities (statistical models).

The notion of statistical manifolds originates from the early foundational works of Rao [Rao45], Chenstov [Cen82] and Amari [Ama85, AN00], which established the notion of Riemannian structure, metric and connnection on statistical models. Due to the dual nature of statistical manifold as both a statistical model and a Riemannian manifold, the geometrical structure of statistical manifolds admits a different characterization compared to Riemannian manifolds studied in "classical" Differential Geometry.

In "classical" Differential Geometry discussed in the previous chapter, the study revolves around intrinsic properties (properties that are invariant under isometry and reparametrization) that reflects the natural characteristics of $\mathbb{R}^n$.

On the other hand, for the geometrical structure on statistical manifolds we would naturally want the geometry to preserve, in addition to the intrinsic properties, the "statistical" properties as well. Therefore we would require a Riemannian structure that is *both* invariant under reparametrization *AND* invariant under sufficient statistics. More formally, the desirable geometrical structure over a statistical model $\{p(x|\theta)\}$ would be invariant under *both* reparametrization of $\theta$ *AND* remapping of $x$ under sufficient statistics of $\theta$.

The above discussion is embodied in a special choice of metric and connection on statistical manifolds that is different from the classical case. In this chapter we

 
R. S. Fong and P. Tino, *Population-Based Optimization on Riemannian Manifolds*, Studies in Computational Intelligence 1046,
https://doi.org/10.1007/978-3-031-04293-5_3

discuss this special type of Riemannian structure summarizing the material discussed in the literature [AN00, CU14, MR93, Lau87, Ama16]. The material in this chapter aims to establish the necessary foundations for the subsequent discussions in the book, and interested readers are refered to the aforementioned literature for detailed derivation and exposition.

This chapter will be organized as follows:

1. In Sect. 3.1, we begin by describing formally the notion of statistical manifolds, its tangent spaces, and the Fisher-Rao Riemannian metric.
2. In Sect. 3.2, we describe the go-to affine connection (Definition 2.16) on Riemannian manifolds in classical Differential Geometry called the Levi-Civita connection. The Levi-Civita connection, otherwise known as Riemannian connection, is the unique affine connection that captures the natural geometrical characteristics of $\mathbb{R}^n$. However, on statistical manifolds a different notion of affine connection arises from the invariance under sufficient statistics. We summarize the historical derivation of $\alpha$-connection from [Daw75, Ama85], and its generalization to dual connections.
3. Finally, in Sect. 3.3, we discuss the notion of curvature and flatness under Levi-Civita connection and the analogous notion of dual flatness under dual connections.

## 3.1 Statistical Manifolds

We begin the discussion by formally introducing the manifold structure of (finitely parametrized) statistical models. In this section we summarize the results from the literature and detail the construction of statistical manifolds formally.

Given a measurable space $M$, consider the space of *all* probability densities on $M$:

$$P(M) = \left\{ p : M \to \mathbb{R} \middle| p \geq 0, \int_M p = 1 \right\} \quad .$$

The space $P(M)$ is an infinite dimensional space, and family of finitely parametrized densities can be view as "finite dimensional slices" of $P(M)$. This is done by immersing a finite dimensional parameter space onto the infinite dimensional $P(M)$, detailed as follows.

Let $\Xi \subset \mathbb{R}^n$ denote a finite dimensional parameter space, and consider the injective immersion:

$$\begin{aligned} \iota : \Xi &\hookrightarrow P(M) \\ \theta = \left(\theta^1, \ldots, \theta^n\right) &\mapsto p_\theta \quad . \end{aligned}$$

Equivalently, this means the map $\iota$ is injective and the set of functions $\left\{\frac{\partial}{\partial\theta^i}p_\theta\right\}_{i=1}^n$ are linearly independent for all $\theta \in \Xi$. A statistical manifold is defined as follows [AN00, CU14]:

**Definition 3.1** An **$n$-dimensional statistical manifold** described by the statistical model (parametric model) $S := \{p_\theta|\theta \in \Xi \subset \mathbb{R}^n\}$ is an $n$-dimensional immersion submanifold of $P(M)$ under the smooth injective immersion $\iota : \Xi \hookrightarrow S = \iota(\Xi) \subset P(M)$. The map $\iota$ is called the **parametrization map** of $S$.

The statistical manifold is invariant under diffeomorphic transformations of parametrization $\iota$. Moreover, since $\iota$ an immersion, the set of linearly independent functions $\left\{\frac{\partial}{\partial\theta^i}p_\theta\right\}_{i=1}^n$ acts as the basis of the tangent spaces of $S$. To see that $\frac{\partial}{\partial\theta^i}p_\theta$ is a derivation (see Definition 2.6), consider a smooth function $f : S \to \mathbb{R}$, then for each $x_0 \in M$ and $i = 1, \ldots n$:

$$\left[\frac{\partial p_\theta}{\partial\theta^i}\right] f\bigg|_{x_0} = \frac{\partial f\,(p(x_0|\theta))}{\partial\theta^i} \in \mathbb{R} \quad .$$

Since Leibniz rule holds for partial derivatives, it also holds for the above expression. Moreover, each element of the basis $\left\{\frac{\partial}{\partial\theta^i}p_\theta\right\}_{i=1}^n$ satisfies:

$$\int_M \frac{\partial}{\partial\theta^i}p(x|\theta)dx = \frac{\partial}{\partial\theta^i}\int_M p(x|\theta)dx = \frac{\partial}{\partial\theta^i}1 = 0 \quad .$$

For any $p \in S$, tangent vectors in $T_pS$ are linear combinations of the above basis vectors, and would therefore also integrate to 0 on $M$. This is an alternative identification of the tangent spaces of statistical manifolds.

To preserve the statistical nature of statistical manifolds $S$, we consider a special basis for each tangent space $T_pS$ given by the **score function** at each $x \in M$:[1]

$$\frac{\partial}{\partial\theta^i}\log p(x|\theta) = \frac{\partial}{\partial\theta^i}\ell_x(\theta) = \partial_i\ell_x(\theta) \quad ,$$

where the last equality is given by the abbreviation $\partial_i = \frac{\partial}{\partial\theta^i}$. Given a smooth function $f : S \to \mathbb{R}$, for each $x_0 \in M$ and $i = 1, \ldots, n$ the above derivation is given by:

$$[\partial_i\ell_x(\theta)]\,f|_{x_0} = \left[\frac{\partial}{\partial\theta^i}\log p(x|\theta)\right] f\bigg|_{x_0} = \left[\frac{1}{p(x_0|\theta)}\frac{\partial p_\theta}{\partial\theta^i}\right] f\bigg|_{x_0} \quad .$$

On statistical manifolds we consider a special Riemannian metric called the **Fisher-Rao information metric**. In local coordinates $(\theta^1, \ldots, \theta^n)$ of $S$, the sym-

[1] For each $i$, the **score function** $\partial_i\ell_x(\theta)$ describes the infinitesimal change of information contained in $p(x|\theta)$ resulting from infinitesimal change in the parameter $\theta^i$. The choice of score functions as a basis of the tangent spaces of statistical manifolds is widely adopted as a convention in the literature of Information Geometry [Rao45, AN00, AJVLS17].

metric positive semi-definite matrix corresponding to the Fisher metric on $S$ (called **Fisher information matrix**) has the following form:

$$F|_\theta = \left[F_{ij}(\theta)\right] = E_\theta\left[\partial_i \ell_x(\theta)\partial_j \ell_x(\theta)\right] = \int_M \partial_i \ell_x(\theta)\partial_j \ell_x(\theta)p(x,\theta)dx \quad ,$$

where $E_\theta[\cdot]$ denote the expected value with respect to $p_\theta$. For the remainder of the book we will adopt the common assumption [AN00] that $F$ is positive definite. Since the score function $\partial_i \ell_x(\theta)$ is invariant under sufficient statistics, so is the Fisher metric. Moreover the Fisher metric is the *unique* Riemannian metric invariant under sufficient statistics up to scalar multiple [Cen82, AJVLS15, VL17].

## 3.2 Levi-Civita Connection and Dual Connections

Using Chrisoffel symbols discussed in the previous chapter, we have (countably) infinite number of choices of connections at our disposal, which are generated simply by choosing a set of $n^3$ functions. There is one connection – the Levi-Civita connection, that uniquely captures the natural geometrical properties on $\mathbb{R}^n$, allowing us to once again inherit the intrinsic properties of Euclidean spaces locally to abstract Riemannian manifolds. In this section we shall discuss the Levi-Civita connection and its characteristics.

Let $(M, g)$ be a Riemannian manifold and $\nabla$ denote an affine connection on $M$. We begin by describing two properties of connections on $M$:

**Definition 3.2** The **torsion tensor** of $\nabla$ is the map $\tau : \mathcal{T}(M) \times \mathcal{T}(M) \to \mathcal{T}(M)$ given by:

$$\tau(X, Y) := \nabla_X Y - \nabla_Y X - [X, Y] \quad ,$$

where $X, Y \in \mathcal{E}(TM)$, and $[X, Y]$ denote the **Lie bracket** $[X, Y] := XY - YX$. An affine connection $\nabla$ is **symmetric** or **torsion-free** if the torsion vanishes identically.

**Definition 3.3** An affine connection is **compatible with** $g$ (or **metric compatible** if the Riemannian metric $g$ is clear) if the following is satisfied for all $X, Y, Z \in T_x M$:

$$\nabla_X \langle Y, Z\rangle = \langle \nabla_X Y, Z\rangle + \langle Y, \nabla_X Z\rangle.$$

The geometrical meaning of these two properties can be thought of as follows:

1. A Riemannian manifold $M$ together with a torsion-free (symmetric) connection means the manifold is not "twisted" along any direction.
2. An affine connection is compatible with metric $\langle\cdot,\cdot\rangle_g$ if and only if the corresponding parallel translation $P_{a,b} : T_a M \to T_b M$ is an isometry for all $a, b \in M$. In other words the following holds for all $X, Y \in T_a M$:

$$\langle X, Y \rangle = \langle P_{a,b}X, P_{a,b}Y \rangle.$$

**Theorem 3.1** (Fundamental Lemma of Riemannian Geometry) *Let $(M, g)$ be a Riemannian manifold, then there exists a unique affine connection $\nabla$ on $M$ that is both compatible with metric $g$ and symmetric. This connection is called the* ***Levi-Civita connection*** *(of $g$).*

For the rest of the discussion, a Riemannian manifold together with its Levi-Civita connection will be written as the triplet: $(M, g, \nabla)$. Given an isometry between Riemannian manifolds, we can always relate their corresponding Levi-Civita connections. The Levi-Civita connection on statistical manifolds corresponding to the Fisher metric is called the **information connection** [Ama85].

**Theorem 3.2** (Naturality of Levi-Civita connection) *Let $(M, g)$ and $\left(\tilde{M}, \tilde{g}\right)$ be Riemannian manifolds with Levi-Civita connections $\nabla, \tilde{\nabla}$ respectively. Suppose $\varphi : (M, g) \to (\tilde{M}, \tilde{g})$ is an isometry then the following is satisfied*

$$\varphi_* \left(\nabla_X Y\right) = \tilde{\nabla}_{\varphi_* X} \left(\varphi_* Y\right) \quad ,$$

*where $X, Y \in \mathcal{E}(TM)$. Equivalently, this means pullback connection $\varphi^* \tilde{\nabla}$ coincides with the Levi-Civita connection $\nabla$ of $M$*[2].

On $n$-dimensional statistical manifolds $S$, we consider an alternative choice of affine connection that is invariant under sufficient statistics: the **$\alpha$-connections**. We summarize the motivation and derivation from [Daw75, Ama85] as follows.

Let $e_i(\theta) := \partial_i \ell_x(\theta)$ for simplicity. Suppose $\delta e_i$ is a sufficiently small $n$-dimensional vector in the ambient space $P(M) \supset S$, such that in local coordinates, when $p_\theta$ and $p_{\theta+d\theta}$ are "close enough", we have: $e_i(\theta + d\theta) = e_i(\theta) + \delta e_i$. Furthermore, we assume $\delta e_i$ depends linearly on the difference in coordinates $d\theta$ when the change is sufficiently small. By the expansion of affine connection in Eq. (2.4), we obtain:

$$e_i(\theta) + \delta e_i = \nabla_{e_i} \sum_{j=1}^{n} d\theta^j e_j = e_i(\theta) + \sum_{j,k=1}^{n} d\theta^k \Gamma_{ij}^k e_k(\theta) \quad .$$

Let $g$ denote the Fisher metric on $S$, the above equation thus gives us:

$$\langle e_m(\theta), \delta e_i \rangle_g = \sum_{j,k=1}^{n} d\theta^j \Gamma_{ij}^k g_{km}(\theta) = \sum_{j=1}^{n} d\theta^j \Gamma_{ijm} \quad ,$$

where $\Gamma_{ijm} = \sum_{k=1}^{n} \Gamma_{ij}^k g_{km}(\theta)$ denote the Christoffel symbol of the first kind.

[2] The pullback $\varphi^*$ of $\varphi$ will be detailed in Sect. 4.1.1.

On the other hand, since $e_i(\theta) = \partial_i \ell_x(\theta)$, when expanding $e_i(\theta + d\theta)$ we obtain:

$$\begin{aligned} e_i(\theta + d\theta) &= \partial_i \ell_x(\theta + d\theta) \\ &= \partial_i \ell_x(\theta) + \sum_{j=1}^{n} \partial_i \partial_j \ell_x(\theta) d\theta^j \quad . \end{aligned}$$

This implies $\delta e_i \cong \sum_{j=1}^{n} \partial_i \partial_j \ell_x(\theta) d\theta^j$. Since Fisher metric is positive definite, $E_\theta \left[ \sum_{j=1}^{n} \partial_i \partial_j \ell_x(\theta) d\theta^j \right] = -g_{ij}(\theta) \neq 0$, the vector $\delta e_i$ does not belong to any tangent space $T_{p_\theta} S$! The author of [Ama85] thus projects the vector $\delta e_i$ onto the tangent space by making the following modification to the connection (equivalently to the parallel transport of vectors):

$$\begin{aligned} \delta^1 &:= \sum_{j=1}^{n} \partial_i \partial_j \ell_x(\theta) d\theta^j + \sum_{j=1}^{n} g_{ij}(\theta)\, d\theta^j \\ \delta^2 &:= \sum_{j=1}^{n} \partial_i \partial_j \ell_x(\theta) d\theta^j + \partial_i \ell_x(\theta) \sum_{j=1}^{n} \partial_j \ell_x(\theta) d\theta^j \\ \delta e_i^{(\alpha)} &:= \frac{1+\alpha}{2} \delta^1 + \frac{1-\alpha}{2} \delta^2 \quad . \end{aligned}$$

By construction both $\delta^1$ and $\delta^2$ have 0 expected value with respect to $\theta$, therefore they both belong to $T_{p_\theta} S$. The **$\alpha$-connection** on $S$, denoted by $\nabla^\alpha$, is therefore defined by the Christoffel symbols of the first kind $\Gamma_{ijm}^{(\alpha)}$ by the following equation:

$$\sum_{j=1}^{n} d\theta^j \Gamma_{ijm}^{(\alpha)}(\theta) = \langle e_m(\theta), \delta e_i^{(\alpha)} \rangle_g = E_\theta \left[ \partial_m \ell_x(\theta) \delta e_i^{(\alpha)} \right]$$

Equivalently, we have the following for each $\theta \in \Xi$:

$$\begin{aligned} \Gamma_{ijm}^{(\alpha)}(\theta) &= E_\theta \left[ \left( \partial_i \partial_j \ell_x(\theta) + \frac{1+\alpha}{2} g_{ij}(\theta) + \frac{1-\alpha}{2} \partial_i \partial_j \ell_x(\theta) \right) \partial_m \ell_x(\theta) \right] \\ \Gamma_{ijm}^{(\alpha)}(\theta) &= E_\theta \left[ \left( \partial_i \partial_j \ell_x(\theta) + \frac{1-\alpha}{2} \partial_i \partial_j \ell_x(\theta) \right) \partial_m \ell_x(\theta) \right] \end{aligned} \tag{3.1}$$

For any $\alpha \in \mathbb{R}$, the $\alpha$-connection is torsion-free (symmetric) and invariant under sufficient statistics[3] [AN00]. Moreover, when $\alpha = 0$ the $\alpha$-connection reduces to the Levi-Civita (information) connection on statistical manifold $S$.

---

[3] This is once again due the fact that the score functions are invariant under sufficient statistics.

In general, whilst $\alpha$-connection is no longer compatible with the metric $g$, it nonetheless satisfies the following more relaxed form:

$$X\langle Y, Z\rangle_g = \langle \nabla^{\alpha}_X Y, Z\rangle_g + \langle Y, \nabla^{(-\alpha)}_X Z\rangle_g \quad , \quad \forall X, Y, Z \in \mathcal{E}(TS) \quad ,$$

where $\mathcal{E}(TS)$ denote the smooth sections of the tangent bundle $TS$, and the corresponding $(-\alpha)$-connection $\nabla^{(-\alpha)}$ is called the **(metric) dual** connection of the $\nabla^{\alpha}$.

More generally and more formally, the notion of $\alpha$-connection and its metric dual $(-\alpha)$-connection is generalized to the notion of a *pair* of dual connections, defined as follows:

**Definition 3.4** Let $(S, g)$ be an $n$-dimensional Riemannian manifold, a pair of affine connections $(\nabla, \nabla^*)$ are **dual** (or $g$**-dual**, $g$**-conjugate**) if the triplet $(g, \nabla, \nabla^*)$ satisfies:

$$X\langle Y, Z\rangle_g = \langle \nabla_X Y, Z\rangle_g + \langle Y, \nabla^*_X Z\rangle_g \quad , \quad \forall X, Y, Z \in \mathcal{E}(TS) \quad ,$$

where $\mathcal{E}(TS)$ denote the smooth sections of the tangent bundle $TS$. The triplet $(g, \nabla, \nabla^*)$ is called the **dualistic structure** on $S$.

In the literature, dualistic structures is alternatively described by the **skewness tensor** (also called **Amari-Chentsov tensor**) [AJVLS17, Ama16] and divergence (also called **contrast function** [CU14, Mat93] (to be discussed in 5.1.1). Furthermore, the notion of dualistic structure also arises in the field of Affine Differential Geometry [NS94] and Hessian manifolds [SY97]. The detailed derivations are beyond the scope of the book, and interested readers are refereed to the literature.

In classical Riemannian manifolds $(M, g)$, Levi-Civita connection is regarded as *the* go-to connection as the natural geometrical properties from Euclidean spaces are preserved. One way to see this by naturality theorem (Theorem 3.2) and Nash embedding theorem [Nas56], where the Levi-Civita coincides nicely with the projected connection (from an ambient Euclidean space) whenever we consider $M$ as a submanifold of some Euclidean space.

However, in the case of statistical manifolds, it would be more natural to consider a pair of dual connections which may no longer be metric compatible instead. To see this we consider the following example:

**Example 3.1** (*Exponential family and* 1*-flatness*) Let $\psi : \Xi \to \mathbb{R}$, and let $\{C, F_1, \ldots F_n\}$ be real valued functions on a measurable space $M$. Suppose $S = \{p_\xi\}$ is the statistical manifold of exponential family on $M$ with elements given by:

$$p_\xi(x) = p(x, \xi) = \exp\left[C(x) + \sum_{i=1}^{n} \xi^i F_i(x) - \psi(\xi)\right]$$

Suppose $\left\{\partial_i = \frac{\partial}{\partial \theta^i}\right\}_{i=1}^{n}$ is a local coordinate frame of $S$, then the 1-connection on $S$ vanishes identically:

$$\begin{aligned}\Gamma^{(1)}_{ij,k} &= E_\xi\left[\partial_i\partial_j\ell_\xi\,\partial_k\ell_\xi\right] \\ &= -\partial_i\partial_j\psi(\xi)\,\underbrace{E_\xi\left[\partial_k\ell_\xi\right]}_{0} = 0 \quad .\end{aligned}$$

From this we see that even though 1-connection is not metric compatible, it does manage to capture the "curvature" of the exponential family $S$ in $P(M)$. In particular, $S$ is "flattened" with respect to 1-connection, and a different set of geometrical tools (such as Pythagorean theorem [AN00]) can thus be inherited from classical Euclidean geometry.

The above example also shows that the exponential family admits a local coordinate frame such that the Chrisoffel symbols of the 1-connection vanish identically, and geodesic curves under 1-connection on $S$ are "flattened" just like straight lines in Euclidean space under the Euclidean connection. In the next section we will discuss the notions of "flatness" and "curvature" formally.

## 3.3 Curvature, Flatness and Dually Flat

A natural way of describing "flatness" of an abstract manifold $M$ is to see whether it resembles a known flat space: $\mathbb{R}^n$. If particular, we say a manifold is flat if for every point there is a local neighbourhood that is *isometric* to an open set in $\mathbb{R}^n$. Following the discussion in [Lee06], consider the Euclidean connection $\nabla$ on $\mathbb{R}^n$: Given vector fields $X, Y, Z$ on the Euclidean space $\mathbb{R}^n$, we would expect the following equality to hold:

$$\nabla_X\nabla_Y Z - \nabla_Y\nabla_X Z = \nabla_{[X,Y]}Z \quad . \tag{3.2}$$

On Riemannian manifold $M$, Levi Civita connection can be viewed as the pullback connection of Euclidean connection from $\mathbb{R}^n$ via the local coordinate chart. Therefore by Theorem 3.2, if a Riemannian manifold $M$ is locally isometric to $\mathbb{R}^n$, then the same condition would hold for Levi-Civita connection on $M$ as well. In particular, we define the curvature tensor to be the measurement of how much the manifold "deviates from flatness".

**Definition 3.5** Let $(M, g, \nabla)$ be a Riemannian manifold, the **curvature** of $\nabla$ is the map $R : \mathcal{T}(M) \times \mathcal{T}(M) \times \mathcal{T}(M) \to \mathcal{T}(M)$ given by:

$$R(X, Y)Z = \nabla_X\nabla_Y Z - \nabla_Y\nabla_X Z - \nabla_{[X,Y]}Z \quad .$$

The **Riemann curvature tensor** is the covariant 4 tensor given by:

$$Rm(X, Y, Z, W) = \langle R(X, Y)Z, W\rangle \quad .$$

Using the above definition, we can show that a manifold is flat if and only if its "deviation from flatness" is zero:

**Theorem 3.3** *A Riemannian manifold $(M, g, \nabla)$ is flat if and only if its curvature vanishes identically.*

Equivalently, Theorem 3.3 implies $(M, g, \nabla)$ is flat if and only if it admits a parallel (local) frame in a neighbourhood around any point. That is, there exists local coordinates $\left(x^1, \ldots, x^n\right)$ and corresponding local coordinate frame $(\partial_i)_{i=1}^n = \left(\frac{\partial}{\partial x^i}\right)_{i=1}^n$ of $M$ such that:

$$\nabla_{\partial_i} \partial_j \equiv 0 \quad .$$

In other words the Christoffel symbols vanish identically. The local coordinate system is called an **affine coordinate system** of $\nabla$.

Moreover, in the case when the manifold $S$ is equipped with a dualistic structure $(g, \nabla, \nabla^*)$, one can show that $S$ is flat with respect to $\nabla$ if and only if $S$ is flat with respect to $\nabla^*$, in particular:

$$\langle R(X, Y)Z, W\rangle = -\langle R^*(X, Y)Z, W\rangle \quad .$$

**Definition 3.6** A manifold $S$ equipped with a torsion-free dualistic structure $(g, \nabla, \nabla^*)$ is **dually flat** if $S$ is flat with respect to $\nabla$ or $\nabla^*$.

Dually flat statistical manifolds encompass a large variety of families of probability distributions including exponential family (1-flat) and mixture family ((−1)-flat) [AN00].

Finally when $(S, g, \nabla, \nabla^*)$ is dually flat, by Theorem 3.3 above, $S$ admits *two* set of local parallel frame corresponding to $\nabla$ and $\nabla^*$ respectively. In particular, there exists two sets of local coordinates $\left(\theta^1, \ldots, \theta^n\right)$ and $(\eta_1, \ldots, \eta_n)$ of $S$ called the **$\nabla$-affine coordinates** and $\nabla^*$ **affine coordinates** respective, such that:

$$\nabla_{\partial_i} \partial_j \equiv 0 \equiv \nabla^*_{\partial^i} \partial^j \quad ,$$

where $\partial_i := \frac{\partial}{\partial \theta^i}$ and $\partial^i := \frac{\partial}{\partial \eta_i}$ for $i = 1, \ldots, n$. The pair of local coordinate $\left(\theta^i\right)_{i=1}^n$ and $(\eta_i)_{i=1}^n$ are called a pair of **dual coordinates** with respect to $g$.

## 3.4 Discussion

This chapter surveys key elements of Information Geometry and summarizes the material described in [AN00, CU14, Ama85, MR93, Lau87, Ama16]. This provides the foundation for the geometrical structure of the *decision space* of the manifold optimization problem in Sect. 1.2. For rest of Part 1 of the book, we discuss the Information Geometrical structure of finitely parametrized statistical models over Riemannian manifolds and establish a geometrical framework for stochastic optimization on Riemannian manifolds.

## References

[AJVLS15] Nihat Ay, Jürgen Jost, Hông Vân Lê, and Lorenz Schwachhöfer. Information geometry and sufficient statistics. *Probability Theory and Related Fields*, 162(1–2):327–364, 2015.

[AJVLS17] Nihat Ay, Jürgen Jost, Hông Vân Lê, and Lorenz Schwachhöfer. *Information Geometry*, volume 64. Springer, 2017.

[Ama85] Shun-ichi Amari. *Differential-Geometrical Methods in Statistics*, volume 28 of *Lecture Notes in Statistics*. Springer, New York, NY, 1985.

[Ama16] Shun-ichi Amari. *Information geometry and its applications*, volume 194. Springer, 2016.

[AN00] S. Amari and H Nagaoka. *Methods of Information Geometry*, volume 191 of *Translations of Mathematical monographs*. Oxford University Press, 2000.

[Cen82] Nikolai Nikolaevich Cencov. *Statistical Decision Rules and Optimal Inference*. Translations of mathematical monographs. American Mathematical Society, 1982.

[CU14] Ovidiu Calin and Constantin Udrişte. *Geometric modeling in probability and statistics*. Springer, 2014.

[Daw75] AP Dawid. Discussion of a paper by bradley efron. *Ann. Stat*, 3(12311234.3), 1975.

[Lau87] Stefan L Lauritzen. Statistical manifolds. *Differential geometry in statistical inference*, 10:163–216, 1987.

[Lee06] John M Lee. *Riemannian manifolds: an introduction to curvature*, volume 176. Springer Science & Business Media, 2006.

[Mat93] Takao Matumoto. Any statistical manifold has a contrast function—on the $c^3$-functions taking the minimum at the diagonal of the product manifold. *Hiroshima Math. J.*, 23(2):327–332, 1993.

[MR93] MK Murray and JW Rice. *Differential Geometry and Statistics*, volume 48. CRC Press, 1993.

[Nas56] John Nash. The imbedding problem for riemannian manifolds. *Annals of mathematics*, pages 20–63, 1956.

[NS94] Katsumi Nomizu and Takeshi Sasaki. *Affine differential geometry: geometry of affine immersions*. Cambridge University Press, 1994.

[Rao45] Calyampudi Radhakrishna Rao. Information and the accuracy attainable in the estimation of statistical parameters. *Bulletin of Calcutta Mathematical Society*, 37:81–91, 1945.

[SY97] Hirohiko Shima and Katsumi Yagi. Geometry of Hessian manifolds. *Differential Geometry and its Applications*, 7(3):277–290, 1997.

[VL17] Hông Vân Lê. The uniqueness of the fisher metric as information metric. *Annals of the Institute of Statistical Mathematics*, 69(4):879–896, 2017.

# Chapter 4
# Probability Densities on Manifolds

**Abstract** This chapter surveys notions of volume form and intrinsic probability distributions on manifolds in the literature, which can be roughly classified into a "geometrical" approach and a "statistical" approach. The "geometrical" approach focuses on the information geometrical structure of the space of *all* probability measures on the base manifolds, whereas the "statistical" approach aims to re-establish point estimation (locally) on manifolds. The chapter begins with a formal description of differential $n$-forms on a manifold, integration on manifolds, and probability density functions on Riemannian manifolds, which provides the reader with the necessary tools for the subsequent discussions. The chapter then describes the "geometrical" approach and the "statistical" approach in detail.

The study of probability densities over manifolds $M$ comes in a variety of fashions originating from different context and purposes. Data measurements on manifolds arise from a wide spectrum of fields, including physics [SR+10], biology [HKK06, KH05], image processing [FJLP03, PS09], computational anatomy [PSP+06] and many more. Depending on the purpose and context, the notion of probability distributions on manifolds can be studied in either an *extrinsic* or *intrinsic* fashion.

Extrinsic constructions of probability densities on include Stiefel manifolds [MK77], parametric submanifolds embedded in $\mathbb{R}^n$ [DHS+13], and algebraic manifolds as subsets of $\mathbb{R}^n$ [BM18]. All these constructions assume additional structures on either the manifold or the ambient space, using additional machinaries such as matrix representation of orthogonal groups, a "global parametrization" of the manifold, pre-defined global algebraic structure, or an immersion of the manifold $M$ from the ambient Euclidean space.

In order to construct manifold optimization algorithms on general manifolds, we therefore turn to the contemporary study of *intrinsic* probability distributions on manifolds in the literature in this chapter.

In the first part of this chapter (Sect. 4.1) we will follow the theme of studying the intrinsic geometry of a manifold through a "linear scaffolding" of vector bundles. We begin by formally describing the notion of volume measurement and differential forms [Lee01]. We then discuss how a notion of "probability volume" on compact subsets of $M$ can be naturally associated to probability density functions on $M$,

 
R. S. Fong and P. Tino, *Population-Based Optimization on Riemannian Manifolds*,
Studies in Computational Intelligence 1046,
https://doi.org/10.1007/978-3-031-04293-5_4

the latter of which is analogous to the "classical" probability density described on Euclidean spaces $\mathbb{R}^n$. For details of the geometry of differential forms on Euclidean spaces and manifolds, we refer to [Lee01, Bac12].

This natural association of differential forms and probability density functions branched into two distinctive ways of studying probability densities on manifolds. In particular, these two branches stem from the "geometrical" view and the practical "statistical" purposes respectively.

In the second part of this chapter (Sect. 4.2) we will review these two approaches in the literature. We discuss how neither approach is suitable for constructing a family of *parametrized* densities over (Riemannian) manifolds for the optimization purposes: the "geometrical" approach is too general and the "statistical" approach too restrictive. In the next chapter we will combine the *essence* of the two approaches, and construct the information geometrical structures of parametrized mixture distributions on $M$.

## 4.1 Volume on Riemannian Manifold in the Literature

Vector bundles on $M$, together with the notion of affine connection, allow us to mount additional structures on abstract manifolds. These structures in turn provide the backbone to construct and study intrinsic properties of the manifold free from an ambient space.

In this section, we build upon the architecture of tangent bundles and discuss volume measurements on Riemannian manifolds.

### *4.1.1 Co-Tangent Bundle*

Given an $n$-dimensional vector space $V$ over $\mathbb{R}$, the space of linear functionals over $V$, denoted by $V^* := \{\omega : V \to \mathbb{R}\}$, is called the **dual space** of $V$.

Let $(E_1, \ldots, E_n)$ be a basis of $V$, then $V^*$ is an $n$-dimensional vector space over $\mathbb{R}$ spanned by the dual basis $\left(e^1, \ldots, e^n\right)$. The set of dual basis satisfies:

$$e^i\left(E_j\right) = \delta^i_j \quad .$$

**Definition 4.1** Let $M$ be a smooth manifold, for each point $x \in M$ the **cotangent space** at $x$ is the dual space of $T_x M$:

$$T^*_x M := (T_x M)^* \quad .$$

The **cotangent bundle** of $M$ is a vector bundle of rank $n$ over $M$ given by:

$$T^*M = \bigsqcup_{x \in M} T_x^*M = \bigcup_{x \in M} \{x\} \times T_x^*M \quad .$$

An element of $T_x^*M$ is called a **tangent covector** and a (smooth) section of $T^*M$ is called a **covector field** or **(smooth) differential** 1-**form**.[1]

To establish integration and eventually volume measurements on manifolds, we consider an important type of tangent covectors induced by smooth real-valued functions on $M$.

**Definition 4.2** Let $M$ be a smooth manifold and let $f \in C^\infty(M)$. For any $x \in M$, the **differential** of $f$, denoted by $df$ is a smooth 1-form on $M$ given by:

$$(df)_x(X_x) = X_x f \quad , \quad \forall X_x \in T_x M \quad .$$

An important example of differentials is the "dual coordinate basis" on $T_x^*M$. Let $(x^1, \ldots, x^n)$ be local coordinates on a local coordinate chart $(U, \varphi)$ of $M$. Since $x^i$ can be realized as the $i$th index of the local coordinate function $\varphi(p) = (x^1(p), \ldots, x^n(p))$, the **local coordinate coframe on** $U$ is the basis of $T_x^*M$ given by the set of differentials $(dx^1|_x, \ldots, dx^n|_x)$ on $T_x^*M$. For any $x \in U \subset M$, a tangent covector $\omega \in T_x^*M$ can be written uniquely as the linear combination:

$$\omega = \sum_{i=1}^{n} \omega_i(p)\, dx^i\big|_x \quad ,$$

where $\omega_i(x) = \omega\left(\frac{\partial}{\partial x^i}\right)_x \in \mathbb{R}$ for $i = 1, \ldots, n$.

This naturally extends to a set of $n$ smooth sections on $T^*M$: For $i = 1, \ldots, n$:

$$dx^i := \frac{\partial}{\partial x^i} : M \to TM$$
$$x \mapsto dx^i\big|_x \quad .$$

The set of 1-forms $(dx^i)_{i=1}^n$ is called a **coordinate coframe**.

Furthermore, just as we can map tangent vectors from one manifold to another through bundle morphism, we can map tangent covectors from one manifold to another as well.

**Definition 4.3** Consider manifolds $M, N$ and their corresponding cotangent bundles $\pi_M : T^*M \to M$, $\pi_N : T^*N \to N$. Given smooth map $f : M \to N$, the **pullback** of $f$, denoted by $f^*$, is the dual (or transpose, adjoint) linear map of the pushforward. For each $x \in M$, the map $f_* : T_x M \to T_{f(x)}N$ is given by:

$$(f^*\omega)(X)\big|_x = \omega(f_*X)|_{f(x)} \quad ,$$

[1] In the sense that it maps one copy of tangent space to $\mathbb{R}$.

where $\omega \in T^*_{f(x)}N$, $X \in T_xM$, and $f^*\omega \in T^*_xM$. Equivalently, the bundle morphism under the following diagram commutes (notice the direction of $f^*$ is opposite of $f$):

$$\begin{array}{ccc} T^*M & \xleftarrow{f^*} & T^*N \\ \pi_M\downarrow & & \downarrow\pi_N \\ M & \xrightarrow{f} & N \end{array}$$

### 4.1.2 Volume and Density Function

In this section we discuss the notion of volume on manifolds, which can be viewed as a "multi-dimensional length measurement". The notion of intrinsic volume on manifolds is defined analogous to integration on Euclidean spaces, where the volume of a set is obtained by a set of (infinitesimally) small "boxes" that partitions it.

Let $M$ be an $n$-dimensional smooth manifold, we return to our scaffolding of tangent bundles and consider for each $x \in M$ the tangent space $T_xM$ centered at $x$. The volume of a parallelepiped spanned by tangent vectors $\{X_1, \ldots, X_n\} \subset T_xM$ can thus be realized by a real-valued map $\omega : (X_1, \ldots, X_n) \mapsto \mathbb{R}$. This in turn defines a real-valued multilinear map on $n$ copies of $T_xM$ (also known as a **covariant** $n$ **tensor**):

$$\omega : \underbrace{T_xM \times \ldots \times T_xM}_{n \text{ copies}} \to \mathbb{R} \quad .$$

When the ambient dimension is $n$ and if $(X_1, \ldots, X_n)$ is linearly dependent, then we would expect the parallelepiped spanned by $(X_1, \ldots, X_n)$ to have zero volume. It can be shown that covariant $n$ tensors $\omega$ satisfying this property must be **alternating**, in other words $\omega$ satisfies:

$$\omega\left(X_1, \ldots, X_a, \ldots, X_b, \ldots, X_n\right) = -\omega\left(X_1, \ldots, X_b, \ldots, X_a, \ldots, X_n\right) \quad .$$

Furthermore, since $M$ is $n$-dimensional, we would naturally expect subspaces with dimension less than $n$ to have zero ($n$-dimensional) volume. In particular, the notion of volume is described by the set of alternating $n$-tensors on $T_xM$ for each point $x$ in $M$, formally defined as follows:

**Definition 4.4** Let $M$ be a smooth $n$-dimensional manifold, the **bundle of alternating (covariant)** $n$**-tensors** on $M$ is the vector bundle:

$$\Omega^n(M) := \bigsqcup_{x \in M} \Omega^n(T_xM) \quad ,$$

where $\Omega^n(T_xM)$ denote the set of alternating $n$-tensors over the vector space $T_xM$ for each $x \in M$. A (smooth) section of $\Omega^n(M)$ is called a **differential $n$-form**

$\Omega^n(M)$ is a 1-dimensional vector bundle over $M$. In particular, for each $x \in M$, every differential $n$-form $\omega \in \Omega^n(M)$ can naturally be associated to a real valued function $f_\omega : M \to \mathbb{R}$. Consider a smooth nonvanishing differential $n$-form $\mu_0 \in \Omega^n(M)$ on $M$ as a reference measure (in the sense of Radon–Nikodym theorem). The aforementioned relationship can be realized in two different fashions:

1. Let $C^\infty(M)$ denote the set of smooth real-valued functions from $M$ to $\mathbb{R}$. The **Hodge Star Operator** is the linear isomorphism $\star$ defined by:

$$\begin{aligned} C^\infty(M) &\to \mathcal{E}\left(\Omega^n(M)\right) \\ \star f &= f\mu_0 \quad . \end{aligned}$$

2. Let $M$ be a compact manifold. Consider the positive differential $n$-forms on $M$ denoted by $\Omega^n_+(M)$ and the set of positive smooth real-valued functions $f : M \to \mathbb{R}_+$ denoted by $C^\infty_+(M)$, . Then there exists a diffeomorphism defined by [BM16, BBM16]:

$$\begin{aligned} R : C^\infty_+(M) &\to \mathcal{E}\left(\Omega^n_+(M)\right) \\ f &\mapsto f^2\mu_0 \quad . \end{aligned} \tag{4.1}$$

The two formulations above depend on the existence of a globally defined smooth nonvanishing differential $n$-form $\mu_0 \in \Omega^n(M)$ as a reference measure, which exists if and only if $M$ is **orientable**. In particular, the reference measure $\mu_0$ provides us with a notion of "orientation" on the manifold, mapping the local coordinate frame on each tangent space continuously to strictly positive or negative real numbers along $x \in M$.[2] In particular, when fixing an orientation (positive or negative) on an orientable manifold $M$, we may choose an **oriented atlas** of $M$ consisting of **oriented coordinate charts** $\{U_\alpha, \varphi_\alpha\}_{\alpha \in \Lambda_M}$ such that:

1. For each $x$ in $M$, the local coordinate frame corresponding to each oriented chart $\left(\frac{\partial}{\partial_i}\Big|_x\right)_{i=1}^n$ is an oriented basis of $T_xM$. In particular when $M$ is positively (negatively) oriented, the local coordinate frame is a positively (negatively) oriented basis of $T_xM$:

$$\mu_0\left(\frac{\partial}{\partial_1}, \ldots, \frac{\partial}{\partial_n}\right) > 0\,(< 0) \quad .$$

2. The Jacobian matrix of the transition functions of the charts have positive determinant in the local oriented coordinates.

The transition function described above is generally called an **orientation-preserving** map. More formally we have the following definition:

---

[2] Note that due to continuity of the reference $n$-form, the range is either $\mathbb{R}_+$ or $\mathbb{R}_-$ and never both.

**Definition 4.5** let $M$, $N$ be oriented manifolds. A local diffeomorphism $f : M \to N$ is **orientation-preserving** if for each $x \in M$, the pushforward of a positivity (resp. negatively) oriented basis of $T_x M$ along $f$ is a positivity (resp. negatively) oriented basis of $T_{f(x)} N$.

Equivalently, a local diffeomorphism is orientation-preserving (**reversing**) if and only if the Jacobian matrix of $f$ with respect to the oriented local coordinates has positive (negative) determinant.

For the rest of the book we will assume $M$ is orientable unless specified otherwise. It is worth noting that this assumption is can be bypassed with an orientation double cover when $M$ is not orientable. However, this is beyond the scope of the book, interested readers are referred to the literature for further detail.

When $M$ is a Riemannian manifold, we may specify a special reference measure.

**Definition 4.6** Let $(M, g)$ be an oriented $n$-dimensional Riemmanian manifold. Let $(E_1, \ldots, E_n)$ denote an orthonormal frame on $M$ such that: for each $x \in M$, the corresponding local coordinate frame $(E_1|_x, \ldots, E_n|_x)$ is a (positively) oriented orthonormal basis for $T_p M$. Then there exists a *unique* smooth orientation form on $M$ called the **Riemannian volume form** $dV_g$ such that:

$$dV_g(E_1, \ldots, E_n) = 1 \quad .$$

On an oriented manifold $M$, we can choose oriented local coordinates $\left(x^1, \ldots, x^n\right)$ such that the corresponding coordinate frame $(\partial_1, \ldots, \partial_n)$ is an orthonormal frame. The Riemannian volume form in the local coordinates can thus be expressed as:

$$dV_g = \sqrt{\left|\det\left(\left[G_{ij}\right]\right)\right|} dx^1 \wedge \cdots \wedge dx^n \quad ,$$

where $\left[G_{ij}\right]$ denote the matrix corresponding to the Riemannian metic in the local coordinates. The symbol $\wedge$ denote the **wedge product** or **exterior product** of differential forms. The wedge product allows us to construct higher dimensional differential forms from lower dimensional ones. In particular, the wedge product of 1-forms $dx^1, \ldots, dx^n$ is the $n$-form $dx^1 \wedge \cdots \wedge dx^n$. Moreover, in any smooth chart with local coordinates $\left(x^1, \ldots, x^n\right)$, any $n$-form $\omega \in \Omega^n(M)$ can be written as:

$$\omega = f_\omega dx^1 \wedge \cdots \wedge dx^n \quad .$$

Integration of differential $n$-forms **with compact support** on manifolds is similar to the 1-dimensional version. In particular, the integration of a compact supported $n$-form can be realized by the pullback from Euclidean spaces.

Let $\omega$ be an $n$-form compactly supported within a single oriented chart $(U, \varphi)$ of $M$. That is, we suppose $\operatorname{supp}(\omega) \subset U \subset M$, then the **integral of $\omega$ over $M$** is defined by:

$$\underbrace{\int_M \omega}_{\text{on } M} = \underbrace{\int_{\varphi(U)} \left(\varphi^{-1^*}\right) \omega}_{\text{on } \mathbb{R}^n} \quad . \tag{4.2}$$

Similar to the 1-dimensional case, the integral of $n$-forms depends on the "direction" of orientation. Let $M, N$ be oriented manifolds and let $f : M \to N$ be a local diffeomorphism, then for $n$-form $\omega$ on $N$:

$$\int_N \omega = \begin{cases} \int_M f^*\omega & \text{if } f \text{ is orientation-preserving} \\ -\int_M f^*\omega & \text{if } f \text{ is orientation-reversing.} \end{cases} \tag{4.3}$$

**Remark 4.1** In the general case when the compact subset supp $(\omega) \subset M$ is covered by not one, but multiple oriented charts in the open cover $\{U_\alpha, \varphi_\alpha\}_{\alpha\in\Lambda_M}$, we may use a **partition of unity subordinate to the open cover**.

Roughly speaking, partition of unity is a set of continuous real-valued functions $\lambda_\alpha : M \to [0, 1]$ supported within each $U_\alpha$ of the open cover. The set of functions in a partition of unity satisfies: for each $x \in M$ the functions add up to one, that is: $\sum_{\alpha\in\Lambda_M} \lambda_\alpha(x) = 1$. In this sense, each function $\lambda_\alpha$ provides a weight of each $U_\alpha$, and the relative weight of the $U_\alpha$'s on each point $x \in M$ are partitioned and distributed.

Given a compactly supported $n$-form $\omega$ covered by finitely many elements of the open cover $\{U_\alpha, \varphi_\alpha\}_{\alpha\in\Lambda_M}$. Let $\{\lambda_\alpha\}_{\alpha\in\Lambda_M}$ be a partition of unity subordinate to the aforementioned open cover, the integral of $\omega$ becomes:

$$\int_M \omega = \sum_{\alpha\in\Lambda_M} \int_M \lambda_\alpha \cdot \omega \quad ,$$

where the expression on the left consists of pulled-back integrals described in Equation (4.2).

The detailed construction of partition of unity is beyond the scope of this work, interested readers are referred to the classical references (such as [Lee01, Lor08]) for differential geometry.

The ability to integrate over compactly supported volume forms allows us to define the notion of volume on Riemannian manifolds:

**Definition 4.7** Let $M$ be a compact Riemannian manifold, the **volume of** $M$ is given by:

$$\text{Vol}(M) = \int_M dV_g \quad .$$

The study of volume of general (subsets of) manifold is a subject of contemporary research, interested readers are refered to [GS79, Gro82, Cha06] and the references therein.

For the rest of the book we will consider $n$-forms that are compactly supported. In particular, we will focus on the subset of positive compactly supported $n$-forms that integrates to 1 over $M$. This set is formally defined as follows:

**Definition 4.8** Let $M$ be a smooth oriented manifold. The set of **probability $n$-forms** on $M$, denoted by $\text{Prob}(M)$, is the subset of $\Omega^n(M)$ satisfying: for all $\omega \in \text{Prob}(M) \subset \Omega^n(M)$:

- $\omega$ is non-negative: for all $x \in M$, $\omega(x) \geq 0$.
- $\omega$ is compactly supported: $\text{supp}(\omega)$ is compact in $M$.
- $\int_M \omega = 1$ .

Given a nonvanishing differential $n$-form $\mu_0 \in \Omega^n(M)$ as our reference measure, the real-valued function $f_\omega : \text{supp}(\omega) \subset M \to \mathbb{R}$ naturally associated to $\omega \in \text{Prob}(M)$ is called the **probability density function of** $\omega \in \text{Prob}(M)$.

Finally, by the discussion above we will henceforth associate the space of (smooth) differential $n$-form $\omega \in \Omega^n(M)$ and the corresponding (smooth) real-valued function $f_\omega$ on $M$ naturally with respect to a reference measure $\mu_0$. That is, the two notions will be used interchangeably.

## 4.2 Intrinsic Probability Densities on Manifolds

In the beginning of the chapter we briefly discussed how the natural association of density functions and differential forms leads to two different approaches of studying probability distributions on manifolds.

The **"geometrical" branch** [KLMP13, BM16, BBM16] stems from the study of probability measure as a differential $n$-form. Motivated by the theoretical information geometrical interpretation of probability densities over Euclidean spaces, it studies the metric, geodesic and other geometrical structure of the vector bundle of *all* probability densities over compact manifolds as an infinite dimensional manifold.[3]

It is worth noting that the definition of $\text{Prob}(M)$ in this book is slightly different from the one discussed in [BM16, BBM16], where the authors assumed $M$ to be **compact** with no restrictions on the support of $\omega \in \Omega^n(M)$. Whereas in our case we assume the $\omega$ to be compactly supported and make *no* additional assumptions on the compact-ness of $M$.

While both constructions ensures the probability $n$-forms are compactly integrable over $M$, this interchanging of assumption is in fact deeply related to the purpose of the construction. Under the natural association of probability density functions and probability $n$-forms, the authors of [BM16, BBM16] study the geometrical structure *entire* space $\text{Prob}(M)$. On the other hand, we wish to study families of *parametrized* probability densities over *general* manifolds $M$.

[3] This is formally known as a Fréchet manifold. Just as finite dimensional manifolds are locally modeled on Euclidean spaces, Fréchet manifolds are locally modeled on Fréchet spaces. [Mic80].

To avoid confusion, we let $\overline{\text{Prob}}\,(M) := \left\{\omega \in \Omega^n\,(M)\,\middle|\,\omega \geq 0,\, \int_M \omega = 1\right\}$ denote the set of probabilty densities on *compact* manifolds $M$ considered in [BM16, BBM16]. As families of finitely parametrized densities can be considered finite-dimensional statistical manifolds, they can be realized as finite-dimensional submanifolds of the space of *all* probability densities $\overline{\text{Prob}}\,(M)$ (an infinite-dimensional manifold). Fixing a reference measure $\mu_0 \in \Omega^n\,(M)$, the space $\overline{\text{Prob}}\,(M)$ can be endowed with the Fisher-Rao metric as the Riemannian metic, uniquely defined under the action of Diff $(M)$ – the set of diffeomorphic automorphisms on $M$:

$$G_{\mu_0}^{FR}\,(\alpha,\beta) = \int_M \frac{\alpha}{\mu_0}\frac{\beta}{\mu_0}\mu_0$$

where $\alpha, \beta \in \overline{\text{Prob}}\,(M)$, and $\frac{\alpha}{\mu_0} := f_\alpha$, $\frac{\beta}{\mu_0} := f_\beta$ denote the probability density associated to $\alpha, \beta$ respectively. Furthermore the authors of [BBM16] show that, when $M$ is compact $n$-dimensional manifold with $n > 1$, the space of positive $n$-forms $\text{Dens}_+\,(M)$ can be equipped with an extension of the metric $G_{\mu_0}^{FR}\,(\alpha,\beta)$. The geometry of $\text{Dens}_+\,(M)$ is subsequently derived in [BM16].

This approach provides us with the geometry of $\overline{\text{Prob}}\,(M)$, which in turn provides us with geometrical tools for algorithm design such as the notion of (natural) gradient. However, this is difficult to use in practice as both the manifold $M$ and the infinite dimensional space $\overline{\text{Prob}}\,(M)$ are 'too large' in the following sense:

1. First of all, the above construction only assumes the base manifolds $M$ to be compact. The entire base manifolds $M$, compact or not, are generally too large in the sense that there is no meaningful way to compare all points pairwise. Unless we make further assumptions on $M$, it is unclear how we could perform statistical point estimations on a global scale.
2. It is also unclear how the local point estimations on $M$ would estimate the parameters of (finite dimensional) submanifolds of $\overline{\text{Prob}}\,(M)$, set of globally defined densities. In particular, as point estimations on manifolds are strictly local, there is no explicit relation between the parametrization of *globally defined* densities in $\overline{\text{Prob}}\,(M)$ and the *local* point estimations on $M$.
3. Moreover, it is also unclear how we can focus on specific families of densities in $\overline{\text{Prob}}\,(M)$. In particular, suppose if we have a desirable finitely parametrized family of densities inheriting specific properties from Euclidean spaces (such as the family of Gaussian distributions which is isotropic), there is no clear way to determine the corresponding embedding in $\overline{\text{Prob}}\,(M)$.

The **"statistical" branch** [Oll93, Pen04, Pen06, BB08], on the other hand, is motivated by statistical analysis on Riemannian manifold. Originating from fields including directional statistics [JM89, Mar75] and shape space analysis [Sma12, LK+93], this branch focuses on application specific types of manifolds such as sphere [Ken82], projective spaces, Lie groups, and matrix manifolds such as Stiefel manifolds [MK77] and Grassmann manifolds [SBBM17].

**Table 4.1** Conversion of arithmetic operation from Euclidean space to normal neighbourhoods of Riemannian manifolds

| | Euclidean spaces | Riemannian manifold |
|---|---|---|
| Subtraction | $y - x =: \mathbf{xy}$ | $\mathbf{xy} = \exp_x^{-1}(y)$ |
| Distance | $\text{dist}(x, y) := \|\|x - y\|\|$ | $\text{dist}(x, y) = \sqrt{\langle \mathbf{xy} \rangle_g}$ |

The motivation of this branch is to translate statistical tools [Ken91] from Euclidean spaces such as estimators, mean, covariance, and higher order moments to the context of Riemannian manifolds. In order to re-establish /recover statistical tools on manifolds, the approach identifies the necessary ingredients of the statistical tools and finds the maximal domain on the manifold for which these tools can be translated over. We illustrate the approach as follows:

Let $\mathcal{X} := (x_1, \ldots, x_k)$ be points in $\mathbb{R}^n$, consider Frechet mean $\mu_F$ and Covariance Cov on Euclidean spaces $\mathbb{R}^n$, given by:

$$\mu_F := \text{argmin}_{y \in M} \frac{1}{n} \sum_{i=1}^{k} \text{dist}(y, x_i)^2$$
$$\text{Cov}(\mathcal{X}, \mathcal{X}) = E\left[(\mathcal{X} - E[\mathcal{X}])(\mathcal{X} - E[\mathcal{X}])^{\top}\right] \quad ,$$

where $E[\mathcal{X}]$ denote the expected value of $\mathcal{X} = (x_1, \ldots, x_k)$. In order to translate these two statistical objects locally to manifolds $M$, it remains to find subsets of $M$ where an analogous notion of "distance" $\text{dist}(y, x_i)$ and "subtraction" $\mathcal{X} - E[\mathcal{X}]$ is established.[4]

In Riemannian manifolds $M$, the natural choice of such a subset is the normal neighourhood $N_x \subset M$ centered around a point $x \in M$ (see Sect. 2.3) where the Riemannian exponential map $\exp_x$ is a local diffeomorphism. The authors [Mar75, Pen06] thus made the following conversion, summarized in Table 4.1 [Pen19].

Therefore the probability densities (or probability $n$-forms) considered in the "statistical" branch are *all* effectively compactly supported *within* a *single* normal neighbourhood.[5] In lieu of this locality, examples of manifolds considered by this approach in the literature are further assumed to be **complete**, where normal neighbourhoods can be "stretched as far as possible".[6] However, completeness is a strong assumption and manifolds in general are *not* complete; for example the

[4] Since the notion of integration over compactly supported probability densities is well-defined on manifolds.

[5] Although the notion of probability densities are defined generally through the Borel algebra of with the manifold $M$ as a measurable space [Sma12], all the computations are confined to a single normal neighbourhoods.

[6] In the sense that, suppose $M$ is a connected and complete Riemannian manifold, and for any $x \in M$ let $U_x \subset T_x M$ denote the pre-image of normal neighbourhood centered at $x$. Then $M$ can be written as the disjoint union $M = \exp_x(U_x) \cup \exp_x(\partial U_x)$. The set $\exp_x(\partial U_x)$ has Hausdroff dimension $n - 1$ [IT01]. This will be further discussed in Secs. 7.1 and 9.1.

manifold $\mathbb{R}^2 \setminus \{0\}$ is not complete. This renders the above construction strictly *local*. In particular, we would not be able to measure points beyond a single normal neighbourhood, and most manifolds are covered by more than one normal neighbourhood or geodesic ball.

## 4.3 Discussion

For the purpose of constructing stochastic optimization algorithms over general (Riemannian) manifolds $M$, we want to describe *parametrized* probability densities beyond a *single* normal neighbourhood. We also want the point estimations on the search space manifold to be meaningfully represented by the parameters of the distributions.

In this chapter we discussed two approaches of constructing intrinsic probability densities on (Riemannian) manifold $M$. We discussed how neither the geometrical nor the statistical approach is appropriate for our purposes, summarized as follows:

1. "Geometrical" approach is too general. There is no clear way to associate the geometry and statistical parameter of a specific family of finitely parametrized probability densities on $M$.
2. "Statistical" approach is too restrictive. Whilst the statistical properties can be locally inherited through Riemannian exponential map, all the computations are confined within a *single* normal neighbourhood in $M$.

Nonetheless these two approach showed us a way forward respectively:

1. The interchangeability of probability density functions and probability $n$-forms via diffeomorphism: When the (subset of) manifold $M$ is compact, there is a diffeomorphism between function space and space of $n$-forms, hence the geometrical structure can be discussed interchangeably (to be shown in Sect. 5.1).
2. Local inheritance of statistical information: Within the normal neighbourhood, the meaning of statistical point estimation is preserved through the Riemannian exponential map.

In the subsequent chapters of the remainder of Part I, we combine the essence of the two approaches. We extend and generalize the "statistical" approach by incorporating the geometrical structure from the "geometrical" approach. In Chap. 5, we show that the Riemannian exponential map (and more generally, orientation-preserving diffeomorphism) can inherit not just the statistical parameter and estimations, but also the information geometrical structure of finitely parametrized probability densities on tangent spaces. These locally inherited densities are then "stitched together" as a mixture in Chap. 6, covering arbitrarily large totally bounded subsets of $M$, well beyond a single normal neighbourhood.

## References


[Bac12] David Bachman. *A geometric approach to differential forms*. Springer Science & Business Media, 2012.

[BB08] Abhishek Bhattacharya and Rabi Bhattacharya. Statistics on riemannian manifolds: asymptotic distribution and curvature. *Proceedings of the American Mathematical Society*, 136(8):2959–2967, 2008.

[BBM16] Martin Bauer, Martins Bruveris, and Peter W Michor. Uniqueness of the fisher–rao metric on the space of smooth densities. *Bulletin of the London Mathematical Society*, 48(3):499–506, 2016.

[BM16] Martins Bruveris and Peter W Michor. Geometry of the Fisher-Rao metric on the space of smooth densities on a compact manifold. arXiv preprint arXiv:1607.04550, 2016.

[BM18] Paul Breiding and Orlando Marigliano. Sampling from the uniform distribution on an algebraic manifold. arXiv preprint arXiv:1810.06271, 2018.

[Cha06] Isaac Chavel. *Riemannian geometry: a modern introduction*, volume 98. Cambridge university press, 2006.

[DHS+13] Persi Diaconis, Susan Holmes, Mehrdad Shahshahani, et al. Sampling from a manifold. In *Advances in modern statistical theory and applications: a Festschrift in honor of Morris L. Eaton*, pages 102–125. Institute of Mathematical Statistics, 2013.

[FJLP03] P Thomas Fletcher, Sarang Joshi, Conglin Lu, and Stephen M Pizer. Gaussian distributions on lie groups and their application to statistical shape analysis. In *Biennial International Conference on Information Processing in Medical Imaging*, pages 450–462. Springer, 2003.

[Gro82] Michael Gromov. Volume and bounded cohomology. *Publications Mathématiques de l'IHÉS*, 56:5–99, 1982.

[GS79] Robert E Greene and Katsuhiro Shiohama. Diffeomorphisms and volume-preserving embeddings of noncompact manifolds. *Transactions of the American Mathematical Society*, 255:403–414, 1979.

[HKK06] Thomas Hamelryck, John T Kent, and Anders Krogh. Sampling realistic protein conformations using local structural bias. *PLoS Computational Biology*, 2(9):e131, 2006.

[IT01] Jin-ichi Itoh and Minoru Tanaka. The lipschitz continuity of the distance function to the cut locus. *Transactions of the American Mathematical Society*, 353(1):21–40, 2001.

[JM89] PE Jupp and KV Mardia. A unified view of the theory of directional statistics, 1975-1988. *International Statistical Review/Revue Internationale de Statistique*, pages 261–294, 1989.

[Ken82] John T Kent. The fisher-bingham distribution on the sphere. *Journal of the Royal Statistical Society: Series B (Methodological)*, 44(1):71–80, 1982.

[Ken91] Wilfrid S Kendall. Convexity and the hemisphere. *Journal of the London Mathematical Society*, 2(3):567–576, 1991.

[KH05] John T Kent and Thomas Hamelryck. Using the fisher-bingham distribution in stochastic models for protein structure. *Quantitative Biology, Shape Analysis, and Wavelets*, 24:57–60, 2005.

[KLMP13] B Khesin, Jonatan Lenells, G Misiołek, and SC Preston. Geometry of diffeomorphism groups, complete integrability and geometric statistics. *Geometric and Functional Analysis*, 23(1):334–366, 2013.

[Lee01] John M Lee. *Introduction to smooth manifolds*. Springer, 2001.

[LK+93] Huiling Le, David G Kendall, et al. The riemannian structure of euclidean shape spaces: a novel environment for statistics. *The Annals of Statistics*, 21(3):1225–1271, 1993.

[Lor08] W Tu Loring. An introduction to manifolds, 2008.

[Mar75] K. V. Mardia. Statistics of directional data. *Journal of the Royal Statistical Society: Series B (Methodological)*, 37(3):349–371, 1975.

[Mic80] Peter W Michor. *Manifolds of differentiable mappings*, volume 3. Birkhauser, 1980.

[MK77] KV Mardia and CG Khatri. Uniform distribution on a stiefel manifold. *Journal of Multivariate Analysis*, 7(3):468–473, 1977.

[Oll93] Josep M Oller. On an intrinsic analysis of statistical estimation. In *Multivariate Analysis: Future Directions 2*, pages 421–437. Elsevier, 1993.

[Pen04] Xavier Pennec. *Probabilities and statistics on Riemannian manifolds: A geometric approach*. PhD thesis, INRIA, 2004.

[Pen06] Xavier Pennec. Intrinsic statistics on riemannian manifolds: Basic tools for geometric measurements. *Journal of Mathematical Imaging and Vision*, 25(1):127, 2006.

[Pen19] Xavier Pennec. Geometric statistics for computational anatomy - overview and recent advances. Workshop on Geometric Processing in Geometry and Learning from Data in 3D and Beyond, IPAM, 2019.

[PS09] Robert Pless and Richard Souvenir. A survey of manifold learning for images. *IPSJ Transactions on Computer Vision and Applications*, 1:83–94, 2009.

[PSP+06] Jean-Marc Peyrat, Maxime Sermesant, Xavier Pennec, Hervé Delingette, Chenyang Xu, Elliot McVeigh, and Nicholas Ayache. Towards a statistical atlas of cardiac fiber structure. In *International Conference on Medical Image Computing and Computer-Assisted Intervention*, pages 297–304. Springer, 2006.

[SBBM17] Salem Said, Lionel Bombrun, Yannick Berthoumieu, and Jonathan H Manton. Riemannian Gaussian distributions on the space of symmetric positive definite matrices. *IEEE Transactions on Information Theory*, 63(4):2153–2170, 2017.

[Sma12] Christopher G Small. *The statistical theory of shape*. Springer Science & Business Media, 2012.

[SR+10] Gabriel Stoltz, Mathias Rousset, et al. *Free energy computations: A mathematical perspective*. World Scientific, 2010.

# Chapter 5
# Dualistic Geometry of Locally Inherited Parametrized Densities on Riemannian Manifolds

**Abstract** The chapter begins by discussing how neither the "geometrical" nor "statistical" approach described in Chap. 4 is suitable for our purpose of establishing a geometrical framework for population-based stochastic optimization on manifolds: the "geometrical" approach is too general whereas the "statistical" approach is too restrictive. We therefore combine the essence of the two branches and develop the notion of locally inherited probability densities on $M$. In particular, we construct probability densities on manifolds geometrically as elements of the density bundle whose parameters are coherent with statistical estimations on manifolds. The proposed framework preserves both the information geometrical structure of volume forms and the statistical meanings of point estimations. This generalizes the "statistical" approach using both the insights of the "geometrical" approach and the machineries of information geometry. However, the locality of the "statistical" approach still persists.

In the next two chapters, we study the Information Geometrical structure of families of intrinsic, (finitely) parametrized probability densities supported on the manifold beyond the normal neighbourhoods, where the statistical parameters are compatible with the point estimations. This establishes the geometrical foundations of an extension framework for stochastic optimization on Riemannian manifold, linking the Riemannian geometry of the search space manifold and the Information Geometry of the decision space. The latter bridges the Information Geometrical structure, statistical parameters and point estimations on the search space manifold beyond the confines of a single normal neighbourhood.

In the previous chapter, we described two branches of constructing intrinsic probability distributions on Riemannian manifolds $M$ in the current literature. Whilst we discussed how neither approach is suitable for our purpose of constructing a geometrical framework for stochastic optimization on Riemannian manifolds, they showed us a way forward.

The "statistical" approach operates on a *single* normal neighbourhood $N_x$ centered at $x \in M$, where computational and statistical tools on $N_x \subset M$ can be translated *locally* from equivalent notions on the tangent space (which is a Euclidean space).



R. S. Fong and P. Tino, *Population-Based Optimization on Riemannian Manifolds*, Studies in Computational Intelligence 1046,
https://doi.org/10.1007/978-3-031-04293-5_5

Local statistical tools and probability distributions are re-established on $M$ within a *single* normal neighbourhood via the Riemannian exponential map.

Under the "geometrical" approach, the notion of probability density functions and probability $n$-forms on $M$ can be viewed interchangably. Furthermore, the space of probability density functions (equivalently, the vector bundle of probabilty $n$-forms) over $M$ can be endowed with a statistical geometrical structure [AN00], which is different from the Riemannian geometry of the base space.

In this chapter we combine the essence of the two approaches. Consider a subset $\tilde{N} \subset M$ with an *orientation-preserving diffeomorphism* $\rho : U \subset \mathbb{R}^n \to \tilde{N} \subset M$. We construct a family of finitely parametrized probability densities $\tilde{S}$ on $\tilde{N}$ as the image of a family of probability densities $S$ on $U$ under the pullback diffeomorphism $\rho^{-1^*} : \text{Prob}\,(U) \to \text{Prob}\left(\tilde{N}\right)$. The statistical geometry of $\rho^{-1^*} S =: \tilde{S} \subset \text{Prob}\left(\tilde{N}\right)$ is then described explicitly by inheriting the statistical geometry of $S \subset \text{Prob}\,(U)$. This process is described by the pullback bundle morphism, summarized in Fig. 5.1.

The rest of this chapter is organized as follows:

1. In Sect. 5.1, we derive the naturality of dualistic geometry between two Hessian-Riemannian manifolds. That is, given a manifold with a predefined dualistic structure $(S, g, \nabla, \nabla^*)$, a smooth manifold $\tilde{S}$ and diffeomorphism $\varphi : \tilde{S} \to S$, we show that a dualistic structure $(\varphi^* g, \varphi^* \nabla, \varphi^* \nabla^*)$ can be induced on $\tilde{S}$ via $\varphi$. We show that the induced dualistic structure and the corresponding divergence

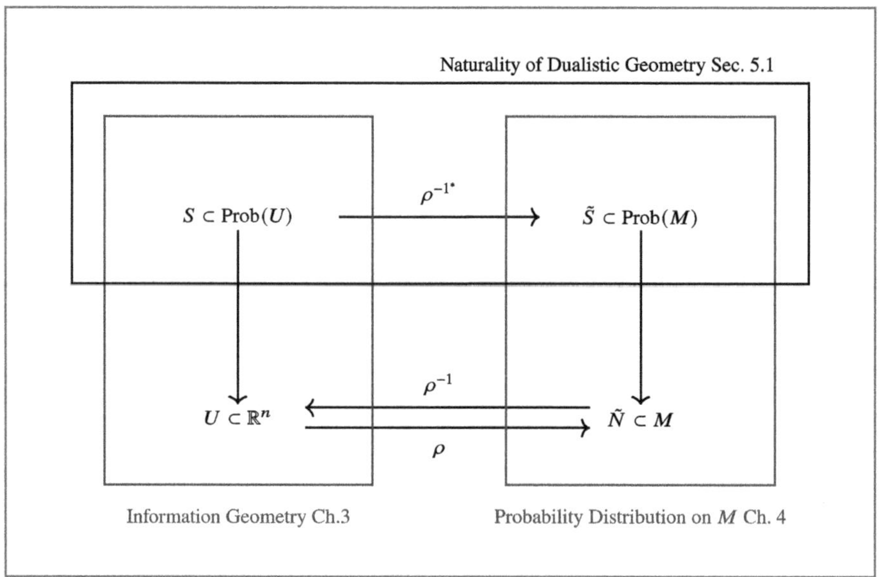


**Fig. 5.1** A summary of the induced statistical dualistic geometry of locally inherited parametrized densities on $M$ described in the book

can be computed explicitly via the pulled-back local coordinates. This describes the top horizontal part of the bundle morphism in Fig. 5.1 boxed in black.
2. In Sect. 5.2, we describe in detail the construction of families of locally inherited probability densities on $M$ summarized in the orientation-preserving bundle morphism with $\varphi := \rho^* : \tilde{S} \to S$ in Fig. 5.1. We discuss the geometrical structure of families of locally inherited probability densities, and show that it generalizes the "statistical" approach described in the previous chapter. This is summarized by the entire bundle morphism of Fig. 5.1 boxed in red.

For the rest of the book we will, without loss of generality, assume $M$ is **connected** and **orientable** unless specified otherwise.

## 5.1 Naturality of Dualistic Structure

Recall from Chap. 3: a statistical manifold $S, g$ is a Riemannian manifold equipped with a pair of $g$-conjugate connections $(\nabla, \nabla^*)$. The triplet $(g, \nabla, \nabla^*)$, called the dualistic structure [AN00], is fundamental to the study of the intrinsic geometry of statistical manifolds [AJVLS17]. The triplet $(g, \nabla, \nabla^*)$ satisfies the following:

$$X\langle Y, Z\rangle_g = \langle \nabla_X Y, Z\rangle_g + \langle Y, \nabla^*_X Z\rangle_g \quad , \quad \forall X, Y, Z \in \mathcal{E}(TS) \quad ,$$

where $\mathcal{E}(TS)$ denote the smooth sections of the tangent bundle $TS$.

In this section we show how dualistic structure of $(S, g, \nabla, \nabla^*)$ can be inherited to an arbitrary smooth manifold $\tilde{S}$ naturally via a diffeomorphism. We show explicitly the relation between induced dualistic structure, Hessian structure and local coordinate systems on finite dimensional statistical manifolds. In particular, we show how one structure can be determined from another computationally. This is represented by the top horizontal (black) part of the bundle morphism in Fig. 5.1.

We discuss two different ways of pulling back (dually flat) dualistic structures given a diffeomorphism from one manifold to another. We first show that general dualistic structures can be pulled back directly via diffeomorphism. We then show when the manifolds are dually flat, the induced dualistic structure can be computed implicitly via the pulled-back coordinates and metric.

Whilst the first method arises more naturally in a theoretical setting, the second provides a more computable way to describe the induced dualistic structure which is equivalent to the first method when the manifold $S$ is dually flat.

We first derive the naturality of dualistic structures, which is an adaptation of naturality of Levi-Civita connection in classical Riemannian geometry (Theorem 3.2). This provides the foundations for the subsequent discussions of the geometry of inherited probabilty densities on manifolds.

Suppose $S$ is a finite dimensional manifold equipped with torsion-free dually flat dualistic structure $(g, \nabla, \nabla^*)$, we can induce a dualistic structure onto another manifold $\tilde{S}$ via a diffeomorphism $\varphi : \tilde{S} \to S$, formally as follows.

**Theorem 5.1** *Let $(S, g, \nabla, \nabla^*)$ be an n-dimensional Riemannian manifold equipped with torsion-free dualistic structure, and let $\tilde{S}$ be an n-dimensional smooth manifold. Suppose $\varphi : \tilde{S} \to S$ is a diffeomorphism, then $\tilde{S}$ can be equipped with an* ***induced torsion-free dualistic structure*** $(\varphi^* g, \varphi^* \nabla, \varphi^* \nabla^*)$.

***Proof*** Let $S$ be a smooth n-dimensional manifold with torsion-free dualistic structure $(g, \nabla, \nabla^*)$, then the following condition is satisfied (Definition 3.4):

$$X\langle Y, Z\rangle_g = \langle \nabla_X Y, Z\rangle_g + \langle Y, \nabla^*_X Z\rangle_g \quad , \quad \forall X, Y, Z \in \mathcal{E}(TS) \quad .$$

Let $\tilde{S}$ be a smooth n-dimensional manifold, and let $\varphi : \tilde{S} \to S$ be a diffeomorphism. The pullback of $g$ along $\varphi$ defines a Riemannian metric on $\tilde{S}$ given by $\tilde{g} = \varphi^* g$. Consider the pullback connection of $\nabla$ on $\tilde{S}$ via $\varphi$ given by:

$$\begin{aligned} \varphi^*\nabla : \mathcal{E}(T\tilde{S}) \times \mathcal{E}(T\tilde{S}) &\to \mathcal{E}(T\tilde{S}) \\ (\varphi^*\nabla)(\tilde{X}, \tilde{Y}) = \varphi_*^{-1}\nabla(\varphi_*\tilde{X}, \varphi_*\tilde{Y}) &= \varphi_*^{-1}\nabla_{\varphi_*\tilde{X}}\varphi_*\tilde{Y} \quad , \end{aligned} \tag{5.1}$$

where $\mathcal{E}(T\tilde{S})$ denote the set of smooth sections of tangent bundle over $\tilde{S}$, $\varphi_*$ denote the push-forward of $\varphi$, and $\varphi^*$ denote the pullback of $\varphi$. Since pullback of torsion-free connection by diffeomorphism is a torsion-free connection, the pullback connections $\varphi^*\nabla$ and $\varphi^*\nabla^*$ are torsion-free connections on the tangent bundle over $\tilde{S}$.

It remains to show that the pair of pullback connections $(\tilde{\nabla}, \tilde{\nabla}^*) := (\varphi^*\nabla, \varphi^*\nabla^*)$ is a $\tilde{g}$-conjugate pair of connections on $\tilde{S}$. In particular, we show that the pair $(\tilde{\nabla}, \tilde{\nabla}^*)$ satisfies the following equation:

$$\tilde{X}\langle \tilde{Y}, \tilde{Z}\rangle_{\tilde{g}} = \langle \tilde{\nabla}_{\tilde{X}}\tilde{Y}, \tilde{Z}\rangle_{\tilde{g}} + \langle \tilde{Y}, \tilde{\nabla}^*_{\tilde{X}}\tilde{Z}\rangle_{\tilde{g}}, \quad \forall \tilde{X}, \tilde{Y}, \tilde{Z} \in \mathcal{E}(T\tilde{S}) \quad .$$

Let $\tilde{X}, \tilde{Y}, \tilde{Z} \in \mathcal{E}(T\tilde{S})$, and let $\tilde{p} \in \tilde{S}$ be an arbitrary point, then:

$$\begin{aligned} \langle \tilde{\nabla}_{\tilde{X}}\tilde{Y}, \tilde{Z}\rangle_{\tilde{g}}(\tilde{p}) &= \langle ((\varphi^*\nabla)_{\tilde{X}}\tilde{Y}, \tilde{Z}\rangle_{\tilde{g}}(\tilde{p}) = \langle \varphi_*^{-1}\nabla_{\varphi_*\tilde{X}}\varphi_*\tilde{Y}_{\tilde{p}}, \tilde{Z}_p\rangle_{\tilde{g}} \\ &= \langle \varphi_*^{-1}\left(\nabla_{\varphi_*\tilde{X}}\varphi_*\tilde{Y}\right)_{\varphi(\tilde{p})}, \varphi_*^{-1}\left(\varphi_*\tilde{Z}\right)_{\varphi(\tilde{p})}\rangle_{\tilde{g}} \\ &= \left(\varphi^{*^{-1}}\tilde{g}\right)_{\varphi(\tilde{p})}\left(\nabla_{\varphi_*\tilde{X}}\varphi_*\tilde{Y}_{\varphi(\tilde{p})}, \varphi_*\tilde{Z}_{\varphi(\tilde{p})}\right) \\ &= \langle \nabla_{\varphi_*\tilde{X}}\varphi_*\tilde{Y}_{\varphi(\tilde{p})}, \varphi_*\tilde{Z}_{\varphi(\tilde{p})}\rangle_g = \langle \nabla_{\varphi_*\tilde{X}}\varphi_*\tilde{Y}, \varphi_*\tilde{Z}\rangle_g(\varphi(\tilde{p})) \quad . \end{aligned}$$

Similarly, by symmetry we obtain the following:

$$\langle \tilde{Y}, \tilde{\nabla}^*_{\tilde{X}}\tilde{Z}\rangle_{\tilde{g}}(\tilde{p}) = \langle \varphi_*\tilde{Y}, \nabla_{\varphi_*\tilde{X}}\varphi_*\tilde{Z}\rangle_g(\varphi(\tilde{p})) \quad .$$

Since $(\nabla, \nabla^*)$ is $g$-conjugate pair of connection on $S$, we have for each $\tilde{p} \in \tilde{S}$:

$$\langle \nabla_{\varphi_*\tilde{X}}\varphi_*\tilde{Y}_{\varphi(\tilde{p})}, \varphi_*\tilde{Z}_{\varphi(\tilde{p})}\rangle_g + \langle \varphi_*\tilde{Y}_{\varphi(\tilde{p})}, \nabla^*_{\varphi_*\tilde{X}}\varphi_*\tilde{Z}_{\varphi(\tilde{p})}\rangle_g = \varphi_*\tilde{X}\left(\langle \varphi_*\tilde{Y}, \varphi_*\tilde{Z}\rangle_g\right)(\varphi(\tilde{p})).$$

Hence on $\varphi(\tilde{p}) \in S$ we have:

$$\begin{aligned}\varphi_* \tilde{X}\left(\langle \varphi_* \tilde{Y}, \varphi_* \tilde{Z}\rangle_g\right)(\varphi(\tilde{p})) &= \varphi_* \tilde{X}\left(\langle \varphi_* \tilde{Y}, \varphi_* \tilde{Z}\rangle_g \circ \varphi \circ \varphi^{-1}\right)(\varphi(\tilde{p})) \\ &= \varphi_* \tilde{X}\left((\varphi^* g)(\tilde{Y}, \tilde{Z}) \circ \varphi^{-1}\right)(\varphi(\tilde{p})) \\ &= \tilde{X}\left(\langle \tilde{Y}, \tilde{Z}\rangle_{\tilde{g}} \circ \varphi^{-1} \circ \varphi\right) \circ \varphi^{-1}(\varphi(\tilde{p})) \\ &= \tilde{X}\langle \tilde{Y}, \tilde{Z}\rangle_{\tilde{g}}(\tilde{p}) \quad .\end{aligned}$$

Combining the above results we obtain the following:

$$\langle \tilde{\nabla}_{\tilde{X}} \tilde{Y}, \tilde{Z}\rangle_{\tilde{g}}(\tilde{p}) + \langle \tilde{Y}, \tilde{\nabla}^*_{\tilde{X}} \tilde{Z}\rangle_{\tilde{g}}(\tilde{p}) = \tilde{X}\langle \tilde{Y}, \tilde{Z}\rangle_{\tilde{g}}(\tilde{p}) \quad , \quad \forall \tilde{X}, \tilde{Y}, \tilde{Z} \in \mathcal{E}(T\tilde{S}) \quad .$$

□

Therefore $\tilde{S}$ can be equipped with the induced torsion-free dualistic structure $(\tilde{g}, \tilde{\nabla}, \tilde{\nabla}^*) = (\varphi^* g, \varphi^* \nabla, \varphi^* \nabla^*)$.

**Remark 5.1** By the proof of the above theorem, the diffeomorphism $\varphi : \left(\tilde{S}, \varphi^* g\right) \to (S, g)$ is an isometry. Furthermore, if $(\nabla, g)$ satisfies Codazzi's equation on $S$ [NS94]:

$$Xg(Y, Z) - g(\nabla_X Y, Z) - g(Y, \nabla_X Z) = (\nabla_X g)(Y, Z) = (\nabla_Z g)(Y, X) \quad ,$$

where $X, Y, Z \in \mathcal{E}(TS)$ denote vector fields on $S$. Then it is immediate that the pair of pulled back connection and metric, denoted by $\left(\tilde{\nabla}, \tilde{g}\right) := (\varphi^* \nabla, \varphi^* g)$, also satisfies Codazzi's equation. Therefore by [Shi07], $\tilde{g}$ is a Hessian metric with respect to $\tilde{\nabla}$. That is, there exists a real-valued **potential** function $\tilde{\psi}$ on $\tilde{S}$ such that $\tilde{g} = \tilde{\nabla} \tilde{\nabla} \tilde{\psi}$, and the triplet $(\tilde{S}, \tilde{g}, \tilde{\psi})$ is called a **Hessian manifold**.

Moreover, if $(g, \nabla, \nabla^*)$ is invariant under sufficient statistics on $S$, then the induced dualistic structure $(\varphi^* g, \varphi^* \nabla, \varphi^* \nabla^*)$ on $\tilde{S}$ is also invariant under sufficient statistics [Cen82].

The pullback curvature on $\left(\tilde{S}, \varphi^* g, \varphi^* \nabla, \varphi^* \nabla^*\right)$ can thus be determined explicitly by an immediate corollary:

**Corollary 5.1** *Let $\varphi : (\tilde{S}, \tilde{g}) \to (S, g)$ be an isometry with $\tilde{g} = \varphi^* g$. Suppose $S$ is equipped with a pair $g$-conjugate connections $(\nabla, \nabla^*)$, and let $(\tilde{\nabla}, \tilde{\nabla}^*) := (\varphi^* \nabla, \varphi^* \nabla^*)$ denote the induced $\tilde{g}$-conjugate connections on $\tilde{S}$. Then the Riemannian curvature tensor $\widetilde{Rm}, \widetilde{Rm}^*$ on $\tilde{S}$ with respect to $\tilde{\nabla}, \tilde{\nabla}^*$ is given by $\widetilde{Rm} = \varphi^* Rm$ and $\widetilde{Rm}^* = \varphi^* Rm^*$ respectively. In particular if $S$ is dually flat, then so is $\tilde{S}$.*

***Proof*** Let $\varphi : (\tilde{S}, \tilde{g}, \tilde{\nabla}, \tilde{\nabla}^*) \to (S, g, \nabla, \nabla^*)$ be a diffeomorphism described in Theorem 5.1, where $(\tilde{g}, \tilde{\nabla}, \tilde{\nabla}^*) := (\varphi^* g, \varphi^* \nabla, \varphi^* \nabla^*)$. Recall Eq. (5.1) in the proof of Theorem 5.1:

$$\tilde{\nabla}_{\tilde{X}}\tilde{Y} = (\varphi^*\nabla)_{\tilde{X}}\tilde{Y} = \varphi_*^{-1}(\nabla_{\varphi_*\tilde{X}}\varphi_*\tilde{Y}), \quad \forall \tilde{X}, \tilde{Y} \in \mathcal{E}\left(T\tilde{S}\right) \quad .$$

By the above equation we have:

$$\begin{aligned} \varphi_*\left(\tilde{\nabla}_{\tilde{X}}\tilde{\nabla}_{\tilde{Y}}\tilde{Z}\right) &= \nabla_{\varphi_*\tilde{X}}\nabla_{\varphi_*\tilde{Y}}\varphi_*\tilde{Z} \quad , \\ \varphi_*\left(\tilde{\nabla}_{\left[\tilde{X},\tilde{Y}\right]}\tilde{Z}\right) &= \nabla_{\varphi_*\left[\tilde{X},\tilde{Y}\right]}\varphi_*\tilde{Z} = \nabla_{\left[\varphi_*\tilde{X},\varphi_*\tilde{Y}\right]}\varphi_*\tilde{Z} \quad . \end{aligned} \tag{5.2}$$

The first equation of Eq. (5.2) is obtained by the following expansion:

$$\begin{aligned} \varphi_*\left(\tilde{\nabla}_{\tilde{X}}\tilde{\nabla}_{\tilde{Y}}\tilde{Z}\right) &= \varphi_*\tilde{\nabla}\left(\tilde{X}, \tilde{\nabla}_{\tilde{Y}}\tilde{Z}\right) = \nabla\left(\varphi_*\tilde{X}, \varphi_*\tilde{\nabla}_{\tilde{Y}}\tilde{Z}\right) \\ &= \nabla\left(\varphi_*\tilde{X}, \nabla\left(\varphi_*\tilde{Y}, \varphi_*\tilde{Z}\right)\right) = \nabla_{\varphi_*\tilde{X}}\nabla_{\varphi_*\tilde{Y}}\varphi_*\tilde{Z} \quad . \end{aligned}$$

Let $\tilde{X}, \tilde{Y}, \tilde{Z}, \tilde{W} \in \mathcal{E}\left(T\tilde{S}\right)$ be vector fields on $\tilde{S}$, the Riemannian curvature tensor $\widetilde{Rm}$ on $\tilde{S}$ with respect to $\tilde{\nabla}$ is thus given by:

$$\begin{aligned} \widetilde{Rm}\left(\tilde{X}, \tilde{Y}, \tilde{Z}, \tilde{W}\right) &= \left\langle \tilde{R}(\tilde{X}, \tilde{Y})\tilde{Z}, \tilde{W}\right\rangle_{\tilde{g}} \\ &= \left\langle \tilde{\nabla}_{\tilde{X}}\tilde{\nabla}_{\tilde{Y}}\tilde{Z}, \tilde{W}\right\rangle_{\tilde{g}} - \left\langle \tilde{\nabla}_{\tilde{Y}}\tilde{\nabla}_{\tilde{X}}\tilde{Z}, \tilde{W}\right\rangle_{\tilde{g}} - \left\langle \tilde{\nabla}_{\left[\tilde{X},\tilde{Y}\right]}\tilde{Z}, \tilde{W}\right\rangle_{\tilde{g}} \\ &= \left\langle \varphi_*^{-1}\left(\nabla_{\varphi_*\tilde{X}}\nabla_{\varphi_*\tilde{Y}}\varphi_*\tilde{Z}\right), \tilde{W}\right\rangle_{\tilde{g}} \\ &\quad - \left\langle \varphi_*^{-1}\left(\nabla_{\varphi_*\tilde{Y}}\nabla_{\varphi_*\tilde{X}}\varphi_*\tilde{Z}\right), \tilde{W}\right\rangle_{\tilde{g}} - \left\langle \varphi_*^{-1}\left(\nabla_{\left[\varphi_*\tilde{X},\varphi_*\tilde{Y}\right]}\varphi_*\tilde{Z}\right), \tilde{W}\right\rangle_{\tilde{g}} \\ &= \left\langle \nabla_{\varphi_*\tilde{X}}\nabla_{\varphi_*\tilde{Y}}\varphi_*\tilde{Z}, \varphi_*\tilde{W}\right\rangle_{g} \\ &\quad - \left\langle \nabla_{\varphi_*\tilde{Y}}\nabla_{\varphi_*\tilde{X}}\varphi_*\tilde{Z}, \varphi_*\tilde{W}\right\rangle_{g} - \left\langle \nabla_{\left[\varphi_*\tilde{X},\varphi_*\tilde{Y}\right]}\varphi_*\tilde{Z}, \varphi_*\tilde{W}\right\rangle_{g} \\ &= \left\langle R(\varphi_*\tilde{X}, \varphi_*\tilde{Y})\varphi_*\tilde{Z}, \varphi_*\tilde{W}\right\rangle_{g} = (\varphi^*Rm)\left(\tilde{X}, \tilde{Y}, \tilde{Z}, \tilde{W}\right) \quad , \end{aligned}$$

where $R$ and $Rm$ denote the curvature and Riemannian curvature tensor on $S$ with respect to $\nabla$ respectively. By symmetry, $\tilde{R}^*$ satisfies:

$$\left\langle \tilde{R}^*(\tilde{X}, \tilde{Y})\tilde{Z}, \tilde{W}\right\rangle_{\tilde{g}} = \left\langle R^*(\varphi_*\tilde{X}, \varphi_*\tilde{Y})\varphi_*\tilde{Z}, \varphi_*\tilde{W}\right\rangle_{g} \quad .$$

If $S$ is dually flat, meaning $R \equiv 0 \equiv R^*$ (by Theorem 3.3), then we have the following equality:

$$R^* \equiv 0 \Leftrightarrow R \equiv 0 \Leftrightarrow \varphi^* R \equiv 0 \Leftrightarrow \varphi^* R^* \equiv 0 \quad .$$

Therefore if $S$ is dually flat, then so is $\tilde{S}$. □

This is a generalization of results on Levi-Civita connection on Riemannian manifold to pairs of dualistic structures.

### 5.1.1 Computing Induced Dualistic Structure

In this section, we discuss how the pulled-back dually flat dualistic structure $(\varphi^* g, \varphi^* \nabla, \varphi^* \nabla^*)$ can be determined explicitly via the pulled-back metric, local coordinates, and the corresponding induced potential function. We begin by reviewing the notion of divergence:

**Definition 5.1** Given a smooth manifold $S$, a **divergence** [AN00] $D$ or **contrast function** [CU14] on $S$ is a smooth function $D : S \times S \to \mathbb{R}_+$ satisfying the following:

1. $D(p; q) \geq 0$, and
2. $D(p, q) = 0$ if and only if $p = q$.

A dualistic structure $\left(g^D, \nabla^D, \nabla^{D^*}\right)$ on $S$ can be constructed from a divergence function $D$ via the following equations [E+92, AN00, CU14], for each point $p, q \in S$:

$$\begin{aligned} g_{ij}^D\big|_p &= g_p^D\left(\partial_i, \partial_j\right) := -\partial_i^1 \partial_j^2 \; D[p; q]|_{q=p} \\ \Gamma_{ijk}^D\big|_p &= \langle \nabla_{\partial_i}^D \partial_j, \partial_k \rangle\big|_p := -\partial_i^1 \partial_j^1 \partial_k^2 \; D[p; q]|_{q=p} \, , \end{aligned}$$

where $(\theta_i)_{i=1}^n$ denote local coordinates about $p \in S$ with corresponding local coordinate frame $(\partial_i)_{i=1}^n$, and $\partial_i^\ell$ denote the $i$th partial derivative on the $\ell$th argument of $D$. By an abuse of notation, we may write [AN00]:

$$\begin{aligned} D[\partial_i; \partial_j] &:= -\partial_i^1 \partial_j^2 \; D[p; q]|_{q=p} \, , \quad \text{and} \\ -D[\partial_i \partial_j; \partial_k] &:= -\partial_i^1 \partial_j^1 \partial_k^2 \; D[p; q]|_{q=p} \quad . \end{aligned} \tag{5.3}$$

**Remark 5.2** Conversely, given a torsion-free dualistic structure and a local coordinate system, there exists a divergence that induces the dualistic structure [Mat93]. We will refer to the divergence $\tilde{D}$ on $\tilde{S}$ corresponding to the induced dualistic structure $(\varphi^* g, \varphi^* \nabla, \varphi^* \nabla^*)$ (not necessarily dually flat) as the **induced divergence** on $\tilde{S}$.

For the rest of the section we will assume both $(S, g, \nabla, \nabla^*)$ and $\left(\tilde{S}, \varphi^* g, \varphi^* \nabla, \varphi^* \nabla^*\right)$ are dually flat $n$-dimensional Riemannian manifolds, where $\tilde{S}$ is equipped with the induced pullback dually flat dualistic structure. We show how the induced pullback dually flat dualistic structure and the corresponding $(\varphi^* \nabla, \varphi^* \nabla^*)$-affine coordinates can be determined explicitly.

**Theorem 5.2** *Let $(S, g, \nabla, \nabla^*)$ be an n-dimensional Riemannian manifold equipped with torsion-free dualistic structure, and let $\left(\tilde{S}, \varphi^* g, \varphi^*\nabla, \varphi^*\nabla^*\right)$ be an n-dimensional Riemannian manifold endowed with pullback dualistic structure via diffeomorphism $\varphi : \tilde{S} \to S$. If S is dually flat, then:*

1. *There exists a pair of $\varphi^* g$-dual affine coordinates with respect to $(\varphi^*\nabla, \varphi^*\nabla^*)$ on $\tilde{S}$, and*
2. *The pair of $(\varphi^*\nabla, \varphi^*\nabla^*)$-affine coordinates on $\tilde{S}$ equals to the pullback of the $(\nabla, \nabla^*)$-affine coordinates on S by $\varphi$ upto an additive constant.*

***Proof*** Let $\left(\partial_i := \frac{\partial}{\partial\theta_i}\right)_{i=1}^n$ denote the local coordinate frame for $TS$ corresponding to local $\nabla$-affine coordinates $(\theta_i)_{i=1}^n$. Consider the tangent subbundle $\mathcal{D} \subseteq T\tilde{S}$ spanned by vector fields $\left(\tilde{\partial}_i := \varphi_*^{-1}\partial_i\right)_{i=1}^n$, then by Eq. 5.1, for each $i, j = 1, \ldots, n$:

$$\begin{aligned}
\left(\varphi^*\nabla\right)_{\varphi_*^{-1}\partial_i} \varphi_*^{-1}\partial_j &= \varphi_*^{-1}\left(\nabla_{\varphi_*\varphi_*^{-1}\partial_i}\varphi_*\varphi_*^{-1}\partial_j\right) \\
&= \varphi_*^{-1}\left(\nabla_{\partial_i}\partial_j\right) = \varphi_*^{-1}(0) = 0
\end{aligned}$$

It remains to show the existence of local coordinates $\left(\tilde{\theta}_i\right)_{i=1}^n$ on $\tilde{S}$ corresponding to the coordinate frame $\left(\tilde{\partial}_i := \varphi_*^{-1}\partial_i\right)_{i=1}^n$, which would be a set of $(\varphi^*\nabla)$-affine coordinates on $\tilde{S}$.

For each $i, j = 1, \ldots, n$, since $\left[\varphi_*^{-1}\partial_i, \varphi_*^{-1}\partial_j\right] = \varphi_*^{-1}\left[\partial_i, \partial_j\right] = 0$ (the zero section), the vector fields $\left(\tilde{\partial}_i := \varphi_*^{-1}\partial_i\right)_{i=1}^n$ commutes. Therefore the tangent subbundle $\mathcal{D} \subseteq T\tilde{S}$ spanned by the vector fields $\left(\tilde{\partial}_i\right)_{i=1}^n$ is involutive. By the theorem of Frobenius, $\mathcal{D}$ is completely integrable, hence there exists local coordinates $\left(\tilde{\theta}_i\right)_{i=1}^n$ on $\tilde{S}$ corresponding to the coordinate frame $\left(\tilde{\partial}_i\right)_{i=1}^n$, The local coordinates $\left(\tilde{\theta}_i\right)_{i=1}^n$ is therefore a set of $(\varphi^*\nabla)$-affine coordinates on $\tilde{S}$.

Let $\left(\partial^i := \frac{\partial}{\partial\eta_i}\right)_{i=1}^n$ denote local coordinate frame of $TS$ corresponding to local $\nabla^*$-affine coordinates $(\eta_i)_{i=1}^n$. By a symmetric argument, there exists local $(\varphi^*\nabla^*)$-affine coordinates $(\tilde{\eta}_i)_{i=1}^n$ on $\tilde{S}$ with corresponding coordinate frame $\left(\tilde{\partial}^i := \varphi_*^{-1}\partial^i\right)_{i=1}^n$. It is then immediate that $\left(\tilde{\theta}_i\right)_{i=1}^n$, $(\tilde{\eta}_i)_{i=1}^n$ are $\varphi^* g$-dual coordinates on $\tilde{S}$: Let $\tilde{g} := \varphi^* g$ denote pullback metric on $\tilde{S}$, then for $i, j = 1, \ldots, n$ we have:

$$\begin{aligned}
\langle\tilde{\partial}_i, \tilde{\partial}^j\rangle_{\tilde{g}} &= \left(\varphi^* g\right)\left(\varphi_*^{-1}\partial_i, \varphi_*^{-1}\partial^j\right) \\
&= g\left(\varphi_*\varphi_*^{-1}\partial_i, \varphi_*\varphi_*^{-1}\partial^j\right) \\
&= g\left(\partial_i, \partial^j\right) = \delta_i^j \quad .
\end{aligned}$$

Finally, consider the pullback of the $\nabla$-affine coordinates $(\theta_i)_{i=1}^n$ of $S$ onto $\tilde{S}$ by $\varphi$ denote by: $\left(\overline{\theta_i} := \theta_i \circ \varphi\right)_{i=1}^n$ and the corresponding coordinate frame $\left(\overline{\partial_i}\right)_{i=1}^n := \left(\frac{\partial}{\partial \overline{\theta_i}}\right)_{i=1}^n$ on $T\tilde{S}$. For $i, j = 1, \ldots, n$:

$$\begin{aligned}\tilde{\partial}_i(\overline{\theta}_j) &= \varphi_*^{-1}\partial_i(\theta_j \circ \varphi) \\ &= \partial_i(\theta_j \circ \varphi \circ \varphi^{-1}) = \partial_i\theta_j = \delta_i^j \quad .\end{aligned}$$

This implies for each $i = 1, \ldots, n$, there exists a constant $c_i$ such that $\tilde{\theta}_i = \overline{\theta}_i + c_i$, which in turn implies $\tilde{\partial}_i = \overline{\partial}_i$ for all $i = 1, \ldots n$. By symmetry, let $\left(\overline{\eta}_i := \eta_i \circ \varphi\right)_{i=1}^n$ denote the pullback coordinates on $\tilde{S}$ of the $\nabla^*$-affine coordinates $(\eta_i)_{i=1}^n$, then there exists constants $d_i$ such that for each $i = 1, \ldots, n$: $\tilde{\eta}_i = \overline{\eta}_i + d_i$.

In other words, the pair of $(\varphi^*\nabla, \varphi^*\nabla^*)$-affine coordinates on $\tilde{S}$ equals to the pullback of the $(\nabla, \nabla^*)$-affine coordinates on $S$ by $\varphi$ upto an additive constant.

□

**Remark 5.3** Without loss of generality, by Theorem 5.2, for the rest of the discussion we will use the $(\varphi^*\nabla, \varphi^*\nabla^*)$-affine coordinates on $\tilde{S}$ and the pullback of the $(\nabla, \nabla^*)$-affine coordinates on $S$ by $\varphi$ interchangeably. It is worth noting that the pullback coordinates and the corresponding local coordinate frame described in this fashion does not depend on the dual flatness of $S$ and $\tilde{S}$. Of course, when $S$ and $\tilde{S}$ are not dually flat, the coordinate systems may no longer be $(\nabla, \nabla^*)$ and $(\varphi^*\nabla, \varphi^*\nabla)$-affine respectively.

Since $\tilde{g}$ is a Hessian metric with respect to $\tilde{\nabla}$, there exists a potential function $\tilde{\psi}$ such that $\tilde{g} = \tilde{\nabla}\tilde{\nabla}\tilde{\psi}$ $\tilde{g}_{ij} = \tilde{\partial}_i\tilde{\partial}_j\tilde{\psi} = \overline{\partial}_i\overline{\partial}_j\tilde{\psi}$. The corresponding $\tilde{g}$-dual local coordinate system of $\tilde{S}$ with respect to $\left(\overline{\theta}_i\right)_{i=1}^n$, denoted by $\left(\overline{\eta}_i\right)_{i=1}^n$ can be defined by $\left(\overline{\eta}_i\right)_{i=1}^n = \left(\overline{\partial}_i\tilde{\psi}\right)_{i=1}^n$ with correspond local coordinate frame $\left(\overline{\partial}^i\right)_{i=1}^n$ of $T\tilde{S}$ [AN00].

We wrap up the section by computing the pullback dualistic structure via the divergence induced by $\tilde{\psi}$ and the pullback dual coordinates.

**Theorem 5.3** *Let $(S, g, \nabla, \nabla^*)$ be an $n$-dimensional Riemannian manifold equipped with torsion-free dualistic structure, and let $\left(\tilde{S}, \tilde{g}, \tilde{\nabla}.\tilde{\nabla}^*\right) := \left(\tilde{S}, \varphi^*g, \varphi^*\nabla, \varphi^*\nabla^*\right)$ be an $n$-dimensional Riemannian manifold endowed with pullback dualistic structure via diffeomorphism $\varphi : \tilde{S} \to S$. Let $\tilde{\psi}$ denote the potential function of the Hessian structure $\left(\tilde{\nabla}, \tilde{g}\right)$. Consider the divergence on $\tilde{S}$ in canonical form corresponding to the $\tilde{g}$-dual coordinates $\left(\overline{\theta}_i\right)_{i=1}^n$ and $\left(\overline{\eta}_i\right)_{i=1}^n$ given by:*

$$\begin{aligned}&\overline{D} : \tilde{S} \times \tilde{S} \to \mathbb{R}_+ \\ &(\tilde{p}, \tilde{q}) \mapsto \overline{D}(\tilde{p}, \tilde{q}) = \tilde{\psi}(\tilde{p}) + \tilde{\psi}^\dagger(\tilde{q}) - \langle \overline{\theta}(\tilde{p}), \overline{\eta}(\tilde{q}) \rangle \quad ,\end{aligned}$$

*where $\tilde{\psi}^\dagger$ is a smooth function on $\tilde{S}$ representing the Legendre-Fréchet transformation of $\tilde{\psi}$ with respect to the pair of $\tilde{g}$-dual local coordinates $\left(\overline{\theta}_i\right)_{i=1}^n$, $\left(\overline{\eta}_i\right)_{i=1}^n$ on $\tilde{S}$. Then:*

*1.* $\left(\tilde{g}, \tilde{\nabla}, \tilde{\nabla}^*\right) = \left(g^{\overline{D}}, \nabla^{\overline{D}}, \nabla^{\overline{D}^*}\right)$, *and*

*2. The Christoffel symbols of* $\left(\tilde{\nabla}, \tilde{\nabla}^*\right)$ *on* $\tilde{S}$ *under the pullback* $\tilde{g}$*-dual coordinate system and the Christoffel symbols of* $(\nabla, \nabla^*)$ *on* $S$ *under the original* $g$*-dual coordinates coincide.*

***Proof*** Let $\overline{g}$ denote the Hessian metric generated by $\overline{D}$: $\overline{g}_{ij}\big|_{\tilde{p}} := g^{\overline{D}}_{ij}\big|_{\tilde{p}}$. By Theorem 5.2, it is immediate that $\tilde{g}_{ij}\big|_{\tilde{p}} = -\overline{\partial}^1_i \overline{\partial}^2_j \overline{D} = g^{\overline{D}}_{ij}\big|_{\tilde{p}}$ for $i, j = 1, \ldots, n$, hence $\tilde{g} = \overline{g}$.

Let $\left(\nabla^{\overline{D}}, \nabla^{\overline{D}^*}\right)$ denote pair of $\tilde{g}$-dual connections defined by $\overline{D}$. We now show $\left(\tilde{\nabla}, \tilde{\nabla}^*\right) = \left(\overline{\nabla}, \overline{\nabla}^*\right) := \left(\nabla^{\overline{D}}, \nabla^{\overline{D}^*}\right)$. Let $\tilde{X}, \tilde{Y}, \tilde{Z} \in \mathcal{E}\left(T\tilde{S}\right)$ and $\tilde{p} \in \tilde{S}$, the following is satisfied by Definition 3.4:

$$\tilde{X}\langle \tilde{Y}, \tilde{Z}\rangle_{\tilde{g}} = \langle \tilde{\nabla}_{\tilde{X}}\tilde{Y}, \tilde{Z}\rangle_{\tilde{g}} + \langle \tilde{Y}, \tilde{\nabla}^*_{\tilde{X}}\tilde{Z}\rangle_{\tilde{g}}$$
$$\tilde{X}\langle \tilde{Y}, \tilde{Z}\rangle_{\overline{g}} = \langle \overline{\nabla}_{\tilde{X}}\tilde{Y}, \tilde{Z}\rangle_{\overline{g}} + \langle \tilde{Y}, \overline{\nabla}^*_{\tilde{X}}\tilde{Z}\rangle_{\overline{g}} \quad .$$

Since $\tilde{g} = \overline{g}$, the two equations are equal. For $p \in \tilde{S}$, let $\left(\overline{\partial}_i = \tilde{\partial}_i\right)_{i=1}^n$ denote the local frame of $T_p\tilde{S}$ corresponding to the $\tilde{\nabla}$-affine corrdinates $\left(\overline{\theta}_i\right)_{i=1}^n$, and let $\tilde{X} = \tilde{\partial}_i = \overline{\partial}_i$, $\tilde{Y} = \tilde{\partial}_j = \overline{\partial}_j$, and $\tilde{Z} = \tilde{\partial}_k = \overline{\partial}_k$ then:

$$\begin{aligned} \langle \tilde{\nabla}_{\tilde{X}}\tilde{Y}, \tilde{Z}\rangle_{\tilde{g}} + \langle \tilde{Y}, \tilde{\nabla}^*_{\tilde{X}}\tilde{Z}\rangle_{\tilde{g}} &= \langle \overline{\nabla}_{\tilde{X}}\tilde{Y}, \tilde{Z}\rangle_{\overline{g}} + \langle \tilde{Y}, \overline{\nabla}^*_{\tilde{X}}\tilde{Z}\rangle_{\overline{g}} \\ = \langle \tilde{\nabla}_{\tilde{\partial}_i}\tilde{\partial}_j, \tilde{\partial}_k\rangle_{\tilde{g}} + \langle \tilde{\partial}_j, \tilde{\nabla}^*_{\tilde{\partial}_i}\tilde{\partial}_k\rangle_{\tilde{g}} &= \langle \overline{\nabla}_{\overline{\partial}_i}\overline{\partial}_j, \overline{\partial}_k\rangle_{\overline{g}} + \langle \overline{\partial}_j, \overline{\nabla}^*_{\overline{\partial}_i}\overline{\partial}_k\rangle_{\overline{g}} \\ = 0 + \langle \tilde{\partial}_j, \tilde{\nabla}^*_{\tilde{\partial}_i}\tilde{\partial}_k\rangle_{\tilde{g}} &= 0 + \langle \overline{\partial}_j, \overline{\nabla}^*_{\overline{\partial}_i}\overline{\partial}_k\rangle_{\overline{g}} \end{aligned}$$

By the proof of Theorem 5.2, since $\tilde{\partial}_i = \overline{\partial}_i$ for all $i$, and $\tilde{g} = \overline{g}$, we have for all $\tilde{p} \in \tilde{S}$:

$$\langle \tilde{\partial}_j, \tilde{\nabla}^*_{\tilde{\partial}_i}\tilde{\partial}_k\rangle_{\tilde{g}} = \langle \tilde{\partial}_j, \overline{\nabla}^*_{\tilde{\partial}_i}\tilde{\partial}_k\rangle_{\tilde{g}} \quad , \quad \forall i, j, k \quad .$$

Therefore $\tilde{\nabla}^* = \overline{\nabla}^*$, and by symmetry we conclude that $\left(\tilde{\nabla}, \tilde{\nabla}^*\right) = \left(\overline{\nabla}, \overline{\nabla}^*\right)$. Furthermore, we can determine the explicit expression of the Christoffel symbols of the induced connection $\overline{\nabla}^* = \tilde{\nabla}^* = \varphi^*\nabla^*$ at $\tilde{p} \in \tilde{S}$ in pullback coordinates $\left(\overline{\theta}_i = \theta_i \circ \varphi\right)_{i=1}^n$.

Let $p \in S$ be an arbitrary, and let $\tilde{p} := \varphi^{-1}(p) \in \tilde{S}$, then the Christoffel symbols coincide:

$$\begin{aligned} \tilde{\Gamma}^*_{ijk}\Big|_{\tilde{p}} &= \overline{\partial}_i\overline{\partial}_j\overline{\partial}_k\tilde{\psi}(\tilde{p}) = \overline{\partial}_i\tilde{g}_{jk}(\tilde{p}) \\ &= \tilde{\partial}_i\tilde{g}_{jk}(\tilde{p}) = \left(\varphi_*^{-1}\partial_i\right)\tilde{g}_{jk}(\tilde{p}) \\ &= \partial_i\left(g_{jk} \circ \varphi \circ \varphi^{-1}\right)(\varphi(\tilde{p})) = \Gamma^*_{ijk}\Big|_p \quad . \end{aligned}$$

By a symmetry argument, we also have :$\tilde{\Gamma}_{ijk}\big|_{\tilde{p}} = \Gamma_{ijk}\big|_{p}$ as claimed.

□

## 5.2 Locally Inherited Probability Densities on Manifolds

In this section, we describe the notion of locally inherited probability densities over smooth manifolds via orientation-preserving bundle morphism. This extends and generalizes the construction of probability distributions on geodesically complete Riemannian manifolds via Riemanian exponential map of the "statistical" approach described in Chap. 4. The discussion in this section is represented by the entire bundle morphism of Fig. 5.1 boxed in red.

In the "statistical" approach described in Sect. 4.2, probability densities and the corresponding statistical properties on geodesically complete Riemannian manifolds are constructed by inheriting probability densities on the tangent space via the Riemannian exponential map.[1]

Let $U_x \subset T_x M$ denote the region where the exponential map is a diffeomorphism. For each $y \in N_x := \exp_x(U_x)$ there exists a unique $v \in U_x$ such that $\exp_x(v) = y$. Given a probability density function $p$ supported on $U_x \subset T_x M$, a probability density function $\tilde{p}$ on $N_x \subset M$ can be defined by:

$$\tilde{p} = \left(\exp_x^{-1}\right)^* p = \log_x^* p \quad ,$$

where $\log_x$ denote the Riemannian logarithm map centered at $x \in M$. Equivalently, for all $y \in N_x = \exp_x(U_x) \subset M$:

$$\tilde{p}(y) = p\left(\exp_x^{-1}(y)\right) = p\left(\log_x(y)\right) \quad . \tag{5.4}$$

Determining the explicit expression for Riemannian exponential map for general Riemannian manifolds could be computationally expensive in practice, as it involves solving the geodesic equation, which is a second order differential equation. Therefore this section aims to find the explicit expression of parametrized probability distributions on manifolds with a more general map. We extend and generalize the above construction in two ways:

1. Finitely parametrized probability densities are locally inherited via any arbitrary *orientation-preserving diffeomorphisms* from subsets of Euclidean spaces instead of Riemannian exponential map on tangent spaces. The orientation-preserving diffeomorphism serves as a *sufficient condition* of inheriting dualistic geometry of family of probability distributions form Euclidean spaces to Riemannian

[1] It is also worth noting that a similar direction has been pursued by related work in [Jer05], where the author described the inheritance of metric through a *single* local coordinate charts. Further inheritance of the whole geometrical structure such as dualistic structure was not discussed.

manifolds. This generalizes the use of exponential map, whose closed form expression might be difficult to determine in practice.
2. Moreover, we can inherit the *entire* statistical geometry from sets of probability densities over subsets of $\mathbb{R}^n$ (or tangent spaces). This allow us to draw correspondence between the statistical parameters of locally inherited probability densities and the statistical estimations of manifold data sets $y_i \in N_x$ along with their corresponding set of vectors in Euclidean spaces (or tangent spaces).

### 5.2.1 Local Probability Densities on Manifolds via Bundle Morphism

In this section, we describe locally inherited family of finitely parametrized probability densities on smooth manifolds induced via an orientation-preserving bundle morphism.

Let $M$ be a smooth topological manifold (not necessarily Riemannian), and let $\text{Prob}(M) \subset \text{Vol}(M)$ denote the vector bundle of probability $n$-forms over $M$ that integrates to 1 described in Chap. 4. Recall that $\text{Prob}(M)$ can be naturally associated to density functions over $M$ given a reference measure $\mu_0$ on $M$.[2] Analogously, we let $\text{Prob}(\mathbb{R}^n)$ denote the volume form over $\mathbb{R}^n$ that integrates to 1.

Let $U$ be a compact subset of $\mathbb{R}^n$, consider a family of finitely parametrized probability density functions supported on $U \subset \mathbb{R}^n$ parametrized by $\theta \in \Xi \subset \mathbb{R}^\ell$: $\hat{S} := \left\{p_\theta : U \to \mathbb{R} \middle| \theta \in \Xi, \int_U p_\theta = 1\right\}$. Without loss of generality, we assume elements of $\hat{S}$ are mutually absolutely continuous, then $\hat{S}$ has the structure of a statistical manifold [AN00].

Let $\mu$ be an arbitrary reference measure on $U \subset \mathbb{R}^n$ (in the sense of Radon-Nikodym theorem), and let $S := \left\{\nu_\theta = p_\theta d\mu \middle| \theta \in \Xi\right\} \subset \text{Prob}(U)$ denote the set of probability $n$-forms over $U$ naturally associated to $\hat{S}$ with respect to $\mu$. Since $U$ is compact, $S$ inherits the dualistic structure of $\hat{S}$ via diffeomorphism $R : C^\infty_+(U) \to \mathcal{E}\left(\Omega^n_+(U)\right)$ in Eq. (4.1) described by the "geometrical" approach in [BBM16].

Consider a map $\rho : U \subset \mathbb{R}^n \to M$ and the corresponding pullback $\rho^{-1^*} : S \subset \text{Prob}(U) \to \text{Prob}(M)$. In order to inherit the dualistic geometry of $S$ to $\rho^{-1^*} S =: \tilde{S} \subset \text{Prob}(M)$, a sufficient condition for $\rho$ is that it must be a diffeomorphism. Moreover, by Equation (4.3), $\rho$ must also be orientation-preserving.

Therefore it suffices to consider an orientation-preserving diffeomorphism $\rho : U \subset \mathbb{R}^n \to M$ and the locally inherited family of probability densities on $M$ induced by $\rho$. In particular, **locally inherited family of probability densities** $\tilde{S} := \rho^{-1^*} S$ **over** $\rho(U) \subset M$ is constructed via the pullback bundle morphism defined by orientation-preserving diffeomorphism[3] $\rho^{-1} : M \to U$:

---

[2] In the case of Riemannian manifold $M$, the reference measure is typically given by the Riemannian volume form: $\mu_0 := dV_g$.

[3] Since $\rho$ is an orientation-preserving diffeomorphism, then so is $\rho^{-1}$.

$$\begin{array}{ccccc} \hat{S} \subset C_+^\infty(U) & \xrightarrow{R} & S \subset \text{Prob}(U) & \xrightarrow{\rho^{-1^*}} & \tilde{S} := \rho^{-1^*} S \subset \text{Prob}(M) \\ & & \downarrow & & \downarrow \\ & & U \subset \mathbb{R}^n & \xleftarrow{\rho^{-1}} & \rho(U) \subset M \end{array} \tag{5.5}$$

Then $\tilde{S} \subset \text{Prob}(M)$ is a family of probability densities over $\rho(U) \subset M$ given by:

$$\tilde{S} := \rho^{-1^*} S = \left\{ \tilde{\nu}_\theta = \rho^{-1^*} \nu_\theta \right\} \quad .$$

More precisely, let $x \in V := \rho(U) \subseteq M$, and let $X_1, \ldots, X_n \in T_x M$ be arbitrary tangent vectors. Given a probability density $\nu_\theta \in S \subset \text{Prob}(U)$, the pullback density $\tilde{\nu}_\theta$ on $V$ is given by:

$$\tilde{\nu}_\theta := \rho^{-1^*} \nu_\theta(X_1, \ldots, X_n)\big|_y = \nu_\theta(\rho_*^{-1} X_1, \ldots, \rho_*^{-1} X_n)\big|_{\rho^{-1}(y)} \quad , \quad \forall y \in V \subseteq M. \tag{5.6}$$

Diagram (5.5) commutes: since $\rho$ is a local diffeomorphism, for each $v \in U$, there exists a unique $y \in \rho(U) \subset M$ such that $y = \rho(v)$, and $(v, p_\theta)$ is a section in the line bundle $\pi_U : \text{Prob}(U) \to U$. We have the following equalities:

$$\begin{aligned} \rho \circ \pi_U(v, p_\theta) &= \rho(v) = y \quad , \\ \pi_M \circ \rho^{-1^*}(v, p_\theta) &= \pi_M \left( \rho^{-1^{-1}}(v), \rho^{-1^*} p_\theta \right) = \rho^{-1^{-1}}(v) = \rho(v) = y \quad , \end{aligned}$$

where $\pi_M : \text{Prob}(M) \to M$ denote the line bundle over $M$.

Moreover, since $\rho$ is an orientation-preserving diffeomorphism, so is $\rho^{-1}$. Therefore we have the following equality on the compact subset $V \subset M$:

$$1 = \int_U \nu_\theta = \int_{V := \rho(U)} \rho^{-1^*} \nu_\theta \quad .$$

Suppose $\nu_\theta \in S$ has probability density function $p_\theta$ with respect to the reference measure $\mu$ on $U$, i.e. $\nu_\theta = p_\theta d\mu$ on $U \subset \mathbb{R}^n$, then in local coordinates $\left(x^1, \ldots, x^n\right)$ of $M$, the above integral can be expressed as follows:

$$1 = \int_U p_\theta d\mu_0 = \int_V \left(p_\theta \circ \rho^{-1}\right) \left(\det D\rho^{-1}\right) dx^1 \wedge \cdots \wedge dx^n \quad ,$$

where $dx^1 \wedge \cdots \wedge dx^n$ denote the reference measure on $M$. Finally, suppose $S$ has dualistic structure given by $(g, \nabla, \nabla^*)$. Since $\rho$ is a diffeomorphism, the pullback $\rho^* : \tilde{S} \to S$ is also a diffeomorphism. Therefore by Theorem 5.1, $\tilde{S}$ can be endowed with an induced dualistic structure $(\varphi^* g, \varphi^* \nabla, \varphi^* \nabla^*)$ with $\varphi := \rho^* = \left(\rho^{-1^*}\right)^{-1}$. In particular, the locally inherited family of probability distributions $\tilde{S} = \rho^{-1^*} S$ inherits the dualistic geometrical structure of $S$ via the pullback bundle morphism.

**Remark 5.4** Note that both $\varphi := \rho^*$ and the local coordinate map $p_\theta \in S \mapsto \theta \in \mathbb{R}^\ell$ are diffeomorphisms. Without loss of generality, for the rest of the book we will write the local coordinate system of $\tilde{S}$ as $(\theta_i)_{i=1}^n$ instead of the pullback local coordinates $\left(\overline{\theta_i} := \theta_i \circ \varphi\right)_{i=1}^n = (\theta_i \circ \rho^*)_{i=1}^n$ for simplicity unless specified otherwise (described by Theorem 5.2 and Remark 5.3) if the context is clear.

**Example 5.1** Suppose $\left(S, g, \nabla^{(\alpha)}, \nabla^{(-\alpha)}\right) \subset \mathrm{Prob}(U)$ is an $\alpha$-affine statistical manifold for some $\alpha \in \mathbb{R}$, with Fisher metric $g$, the associated $g$-dual $\alpha$-connections $\left(\nabla^{(\alpha)}, \nabla^{(-\alpha)}\right)$, and the corresponding $\alpha$-divergence $D_\alpha$ on $S$ [AN00]. Since $\rho : U \to M$ is a (local) diffeomorphism, it is injective, hence a sufficient statistic for $S$ [AN00].

By the invariance of Fisher metric and $\alpha$-connection under sufficient statistic, the induced family $\tilde{S}$ is also an $\alpha$-affine statistical manifold.

Furthermore, due to the monotonicity of $\alpha$-divergence (as a special case of $f$-divergence), the induced divergence $\tilde{D}_\alpha$ (see Remark 5.2) on $\tilde{S}$ can be computed by:

$$\tilde{D}_\alpha(\rho^{-1^*} p, \rho^{-1^*} q) = D_\alpha(p, q) \quad , \quad \text{for } p, q \in S \quad .$$

### 5.2.2 Special Case—"Statistical" Approach of Local Probability Densities on Manifolds via Riemannian Exponential Map

We conclude the chapter with an by illustrating how the above framework generalizes and encapsulates the "statistical" approach described in Sect. 4.2. That is, we show that the "statistical" approach of constructing local probability densities via Riemannian exponential map on complete Riemannian manifolds described in [Pen04] is a special case of the bundle morphism described in diagram (5.5).

**Example 5.2** Let $M$ be a (complete) Riemannian manifold. For each $x \in M$, let $U_x \subset T_x M$ denote the region where the Riemannian exponential map $\exp_x : T_x M \to M$ is a local diffeomorphism onto $N_x := \exp_x(U_x) \subset M$.

Since each tangent $T_x M$ is a topological vector space, it can be considered naturally as a metric space with the metric topology induced by the Riemannian metric. Since finite dimensional topological vector spaces of the same dimension $n := \dim(M)$ are unique up to isomorphism, $T_x M$ is isomorphic to $\mathbb{R}^n$. Moreover, since the Euclidean metric and Riemannian metric are equivalent metrics on finite dimensional topological vector spaces, the respective induced metric topologies are also equivalent. This means probability density functions over $T_x M$ can be considered naturally as density functions over $\mathbb{R}^n$ [Lee01, Pet06].

Let $S_x$ denote a finitely parametrized family of probability densities (equivalently probability $n$-forms) over $U_x$. For each $x \in M$, the pushforward of $\exp_x$, denoted by $(\exp_x)_*$, is the identity map. This implies $\exp_x$ is an orientation preserving diffeomorphism on $U_x$. By replace orientation-preserving diffeomorphism $\rho : U \subset \mathbb{R}^n \to \rho(U) \subset M$ with the Riemannian exponential map $\exp_x : U_x \subset T_xM \to N_x \subset M$, the orientation-preserving bundle morphism of diagram (5.5) reduces to:

$$\begin{array}{ccc} S_x \subset \text{Prob}(U_x) & \xrightarrow{\log_x^*} & \tilde{S}_x \subset \text{Prob}(N_x) \\ \downarrow & & \downarrow \\ U_x \subset T_xM & \xleftarrow{\log_x} & N_x \subset M \end{array} \tag{5.7}$$

where $\log_x^*$ denote the pullback of the Riemannian logarithm function $\log_x = \exp_x^{-1}$ on $N_x$. For $p(\cdot|\theta) \in S_x$, the inherited probability density $\tilde{p}(\cdot|\theta) \in \tilde{S}_x$ over $M$ is given by the following:

$$\tilde{p}(y|\theta) = \log_x^* p(y|\theta) = p(\exp_x^{-1}(y)|\theta) = p(\log_x(y)|\theta) \quad .$$

This coincide with Equation (5.4), and the "statistical" approach described in Sect. 4.2 is therefore a special case of the construction of locally inherited densities via orientation-preserving bundle morphism summarized in diagram (5.5) of Sect. 5.2.1.

It is important to note that for general Riemannian manifolds, this approach maybe quite limiting since $U_x$ maybe a small region in the tangent space (bounded by the injectivity radius) (see Sect. 8.1.1).

**Remark 5.5** Throughout the rest of the book, we will use Riemannian exponential map as an example to illustrate our approach. It is however worth noting that our generalized construction applies to all orientation-preserving diffeomorphisms, not just the Riemannian exponential map.

## 5.3 Discussion and Outlook

Even though the framework of locally inherited densities described in this chapter generalized the "statistical" approach with a geometrical view, the local restriction of the "statistical" approach still persists. Given an orientation preserving diffeomorphism $\rho : U \to M$, the locally inherited densities are only defined within $\rho(U)$. In the case when $\rho : \mathbb{R}^n \to M$ is the Riemannian exponential map, the construction is confined within a single normal neighhbourhood.

In the subsequent chapter, we extend this framework to construct probability distributions on $M$ supported beyond the normal neighbourhood. In particular, we extend the notion of locally inherited parametrized families of probability densities to mixture densities on totally bounded subsets of $M$.

## References

[AJVLS17] Nihat Ay, Jürgen Jost, Hông Vân Lê, and Lorenz Schwachhöfer. *Information Geometry*, volume 64. Springer, 2017.

[AN00] S. Amari and H Nagaoka. *Methods of Information Geometry*, volume 191 of *Translations of Mathematical monographs*. Oxford University Press, 2000.

[BBM16] Martin Bauer, Martins Bruveris, and Peter W Michor. Uniqueness of the fisher–rao metric on the space of smooth densities. *Bulletin of the London Mathematical Society*, 48(3):499–506, 2016.

[Cen82] Nikolai Nikolaevich Cencov. *Statistical Decision Rules and Optimal Inference*. Translations of mathematical monographs. American Mathematical Society, 1982.

[CU14] Ovidiu Calin and Constantin Udrişte. *Geometric modeling in probability and statistics*. Springer, 2014.

[E+92] Shinto Eguchi et al. Geometry of minimum contrast. *Hiroshima Mathematical Journal*, 22(3):631–647, 1992.

[Jer05] IH Jermyn. Invariant bayesian estimation on manifolds. *Ann. Stat.*, 33(math. ST/0506296):583–605, 2005.

[Lee01] John M Lee. *Introduction to smooth manifolds*. Springer, 2001.

[Mat93] Takao Matumoto. Any statistical manifold has a contrast function—on the $c^3$-functions taking the minimum at the diagonal of the product manifold. *Hiroshima Math. J.*, 23(2):327–332, 1993.

[NS94] Katsumi Nomizu and Takeshi Sasaki. *Affine differential geometry: geometry of affine immersions*. Cambridge University Press, 1994.

[Pen04] Xavier Pennec. *Probabilities and statistics on Riemannian manifolds: A geometric approach*. PhD thesis, INRIA, 2004.

[Pet06] Peter Petersen. *Riemannian geometry*, volume 171. Springer, 2006.

[Shi07] Hirohiko Shima. *The geometry of Hessian structures*. World Scientific, 2007.

# Chapter 6
# Mixture Densities on Totally Bounded Subsets of Riemannian Manifolds

**Abstract** In this chapter, the notion of parametrized probability densities over manifolds is extended beyond the confines of a single normal neighbourhood, overcoming the locality of the "statistical" approach described in the previous chapter. In particular, we describe the information geometrical structure of mixture densities over totally bounded subsets of manifolds. This chapter derives the product statistical Riemannian structure of family of mixtures densities, which is *separate* from the manifold structure of the base search space. The components of the statistical geometry such as dualistic structure and canonical divergence are also derived from first principles. This provides us with a computable parametric probability model over arbitrarily large (totally bounded) subsets of Riemannian manifolds, and establishes a geometrical framework for population-based stochastic optimization and estimation over manifolds in the second part of the book.

In the previous chapter, we discussed how parametrized family of probability densities over Riemannian manifolds $M$ can be locally inherited from $\mathbb{R}^n$ via an orientation-preserving diffeomorphism $\rho : U \subset \mathbb{R}^n \to M$. This provides us with a notion of computable, parametrized probability densities over subsets of Riemannian manifolds, generalizing the "statistical" approach described in [Oll93, Pen04, BB08] and Sects. 4.2 and 5.2.2. However, the locality of the "statistical" approach still persisted. In particular, the locally inherited densities are *only* defined within the subset $\rho(U) \subset M$ (analogously a normal neighbourhod, see Remark 5.5).

In this chapter we extend the notion of parametrized probability densities over manifolds beyond the confines of a single normal neighbourhood. In particular, we describe the statistical geometrical structure of mixture densities over totally bounded subsets $V$ of Riemannian manifolds.

Given a totally bounded subset $V$ of $M$, we can cover it by a finite number of metric balls (geodesic balls), where each metric ball is mapped to a subset of $\mathbb{R}^n$ via an orientation-preserving diffeomorphism. This set of metric balls (together with the mapping to $\mathbb{R}^n$) forms an orientation-preserving open cover of $V$.

Suppose each element of the open cover of $V$ is equipped with a family of locally inherited probability distributions, a family of parametrized mixture densities $\mathcal{L}_V$ on the *entire* $V$ can be constructed by "gluing" together the locally inherited densities

 
R. S. Fong and P. Tino, *Population-Based Optimization on Riemannian Manifolds*, Studies in Computational Intelligence 1046,
https://doi.org/10.1007/978-3-031-04293-5_6

in the form of a mixture. The statistical geometry of the family of mixture densities on $V$ can thus be described by the product of the mixture component families and the mixture coefficients.

The rest of the chapter is outlined as follows:

1. In Sect. 6.1, we describe the notion of orientation-preserving open cover on Riemannian manifolds $M$. We show that orientation-preserving open covers admits a refinement of geodesic balls, which implies totally bounded subsets $V$ of $M$ admits a finite orientation-preserving open cover of (compact) geodesic balls.
2. In Sect. 6.2, we describe finitely parametrized mixture distributions $\mathcal{L}_V$ over totally bounded subsets $V$ on Riemannian manifolds $M$. The parametrized mixture probability densities extends the notion of locally inherited probability densities beyond a single normal neighbourhood/geodesic ball, while preserves the locally inherited dualistic statistical geometry of the mixture components.
3. From Sect. 6.3 onwards, we describe the geometrical structure of the family of mixture densities $\mathcal{L}_V$. By viewing the closure of the simplex of mixture coeffieients and family of mixture component densities as statistical manifolds, we show that $\mathcal{L}_V$ is a smooth manifold under two condition. Furthermore, we show that a torsion-free dualistic structure can be defined on $\mathcal{L}_V$ by a mixture divergence $D$ from the statistical manifolds of mixture coefficient and mixture components.
4. In Sect. 6.4, we show that $\mathcal{L}_V$ is in fact a product Riemannian manifold of the simplex of mixture coefficients and the families of mixture component statistical manifolds. Finally, we show that when both the simplex of mixture coefficients and mixture component statistical manifolds are dually flat, then $\mathcal{L}_V$ is also dually flat with canonical mixture divergence $D$.

## 6.1 Refinement of Orientation-Preserving Open Cover

We begin by describing formally the notion of orientation-preserving open cover:

**Definition 6.1** Let $(M, g)$ be a smooth $n$-dimensional Riemannian manifold, an **orientation-preserving open cover** of $M$ is an at-most countable set of pairs $\mathcal{E}_M := \{(\rho_\alpha, U_\alpha)\}_{\alpha \in \Lambda_M}$ satisfying:

1. $U_\alpha \subset \mathbb{R}^n$, $\rho_\alpha : U_\alpha \to M$ are local orientation preserving diffeomorphisms for each $\alpha \in \Lambda_M$,
2. the set $\{\rho_\alpha(U_\alpha)\}_{\alpha \in \Lambda_M}$ is an open cover of $M$.

**Example 6.1** One notable example of an orientation-preserving open cover is given by geodesic balls: Using the notation of Sect. 2.3.1: for $x \in M$, let $B_x := B(\mathbf{0}, \text{inj}(x)) \subset T_x M$ denote the geodesic ball of injectivity radius. Riemannian exponential map $\exp_x : B_x \subset T_x M \to M$ is an orientation-preserving diffeomorphism within $B_x$, hence $\left\{(\exp_x, B_x)\right\}_{x \in M}$ is an orientation-preserving open cover of $M$.

In particular, for any positive $j_x \leq \text{inj}(x)$, the set $\left\{\left(\exp_x, B(\mathbf{0}, j_x)\right)\right\}_{x\in M}$ is also an orientation-preserving open cover of $M$.

**Remark 6.1** It is important to note that the notion of orientation-preserving open cover differs from orientable atlas in the literature, as we do *not* require the transition maps between elements of the open cover to be (smoothly) compatible.

Given an open cover over Riemannian manifold $M$, there exists a refinement of open cover by metric balls (geodesic balls) in $M$. This will be used to construct a finite orientation-preserving open cover of totally bounded subsets of $M$.

**Lemma 6.1** *Let $M$ be a $n$-dimensional Riemannian manifold. Given an open cover $\{(\rho_\alpha, U_\alpha)\}_{\alpha\in\Lambda_M}$ of $M$, there always exist a refinement $\left\{\left(\rho_\beta, W_\beta\right)\right\}_{\beta\in\Lambda'_M}$ of $\{(\rho_\alpha, U_\alpha)\}_{\alpha\in\Lambda_M}$ satisfying:*

1. *$\left\{\rho_\beta\left(W_\beta\right)\right\}_{\beta\in\Lambda'_M}$ covers $M$, and*
2. *$\rho_\beta(W_\beta)$ are metric balls in $M$ for all $\beta \in \Lambda'_M$.*

***Proof*** Given open cover $\{(\rho_\alpha, U_\alpha)\}_{\alpha\in\Lambda_M}$ of $M$. For each $\alpha$, for any point $x_\alpha^\beta \in \rho_\alpha(U_\alpha)$ there exists a normal neighbourhood with $N_{x_\alpha^\beta} \subset \rho_\alpha(U_\alpha)$.

Since $N_{x_\alpha^\beta}$ is open for all $x_\alpha^\beta \in \rho_\alpha(U_\alpha)$, there exists $\epsilon_{x_\alpha^\beta} > 0$ such that the metric ball $B\left(x_\alpha^\beta, \epsilon_{x_\alpha^\beta}\right)$ centred at $x_\alpha^\beta$ satisfies: $B\left(x_\alpha^\beta, \epsilon_{x_\alpha^\beta}\right) \subset N_{x_\alpha^\beta} \subset \rho_\alpha\left(U_\alpha\right) \subset M$. Moreover, $B_{x_\alpha^\beta} := B\left(x_\alpha^\beta, \epsilon_{x_\alpha^\beta}\right)$ is the metric ball centred at $x_\alpha^\beta$ in normal coordinates under the radial distance function.

Since $\rho_\alpha$ is a diffeomorphism for all $\alpha \in \Lambda_M$, this implies for each point $x_\alpha^\beta \in \rho_\alpha(U_\alpha)$ we have: $\rho_\alpha^{-1}(B_{x_\alpha^\beta}) =: W_{x_\alpha^\beta} \subset U_\alpha$. Hence $\left\{\left(\rho_\alpha, W_{x_\alpha^\beta}\right)\right\}_{\beta\in\Lambda'_M}$ is the desired refinement of $\{(\rho_\alpha, U_\alpha)\}_{\alpha\in\Lambda_M}$.

□

Observe that the proof does not require $\rho_\alpha$ to be orientation-preserving, as the transition maps between elements of the open cover *do not* have to be orientation-preserving. In general, it suffices to consider open cover $\{(\rho_\alpha, U_\alpha)\}_{\alpha\in\Lambda_M}$ of $M$ such that $\rho_\alpha$ are just diffeomorphisms with the following result.

**Lemma 6.2** *Let $f : M \to N$ be a local diffeomorphism between manifolds $M, N$, then there exists local orientation-preserving diffeomorphism $\tilde{f} : M \to N$.*

***Proof*** Since $f : M \to N$ is a local diffeomorphism, the pushforward $f_* : T_pM \to T_{f(p)}N$ is a linear isomorphism for all $x \in M$. This implies the determinant of the matrix $Df$ is non-zero: $\det Df \neq 0$. If $f$ is orientation-preserving, then there's nothing left to prove. Hence we will now assume $f$ is orientation reversing, i.e. $\det Df < 0$.

Let $\left(x^1, \ldots, x^n\right)$ denote local coordinates on $M$, and let $\overline{f}$ denote coordinate representation of $f$, then for $p \in M$:

$$\overline{f}\left(x^1(p), \ldots, x^n(p)\right) = \left(\overline{f}_1(x), \ldots, \overline{f}_a(x), \overline{f}_{a+1}(x), \ldots, \overline{f}_n(x)\right),$$

where $x := \left(x^1(p), \ldots, x^n(p)\right)$ is the coordinate representation of $p$.

Choose $a \in [1, \ldots, n]$, and let $\tilde{f} : M \to N$ denote the diffeomorphism from $M$ to $N$ defined by the following coordinate representation:

$$\tilde{f}\left(x^1(p), \ldots, x^n(p)\right) = \left(\overline{f}_1(x), \ldots, \overline{f}_{a+1}(x), \overline{f}_a(x), \ldots, \overline{f}_n(x)\right).$$

In other words, we define $\tilde{f}$ by swapping the $a^{th}$ and $a+1^{th}$ coordinates of $f$. The matrix representation of $\tilde{f}_*$ in standard coordinates is thus given by:

$$D\tilde{f} = I' \cdot Df. \tag{6.1}$$

The matrix $I'$ in Eq. (6.1) is given by:

$$I' = \begin{bmatrix} I_{a-1} & & \\ & \begin{matrix} 0 & 1 \\ 1 & 0 \end{matrix} & \\ & & I_{n-(a+1)} \end{bmatrix},$$

where $I_k$ is the identity matrix of the dimension $k = a-1, n-(a+1)$, the sub-matrix $\begin{bmatrix} 0 & 1 \\ 1 & 0 \end{bmatrix}$ is located at the $(a, a)^{th}$ to the $(a+1, a+1)^{th}$ position of $I'$, and the rest of the entries are all zero. Therefore $\tilde{f} : M \to N$ is the desired orientation-preserving diffeomorphism since $\det Df < 0$, which implies $\det D\tilde{f} = \det I' \cdot \det Df = -1 \cdot \det Df > 0$.

□

For the rest of the discussion we will consider the orientation-preserving open cover by metric balls $\left\{\left(\rho_\alpha, W_{x_\alpha^\beta}\right)\right\}_{\beta \in \Lambda'_M}$ of $n$-dimensional Riemannian manifold $M$.

## 6.2 Mixture Densities on Totally Bounded Subsets of Riemannian Manifolds

In this section, we construct family of parametrized mixture probability densities on totally bounded subsets of Riemannian manifold $M$. Let $V \subset M$ be totally bounded subset of $M$, and let $\{(\rho_\alpha, U_\alpha)\}_{\alpha \in \Lambda_M}$ be an orientation-preserving open cover of $M$. Let $\left\{\left(\rho_\alpha, W_{x_\alpha^\beta}\right)\right\}_{\beta \in \Lambda'_M}$ denote a refinement of $\{(\rho_\alpha, U_\alpha)\}_{\alpha \in \Lambda_M}$ by open metric balls in $M$ described by Lemma 6.1. Since $\left\{\rho_\alpha\left(W_{x_\alpha^\beta}\right)\right\}_{\beta \in \Lambda'_M}$ is an open cover of $M$ by metric balls, it is an open cover of $V \subset M$ as well. Moreover, since $V$ is totally bounded, there exists a finite subcover $\left\{\rho_\alpha\left(W_{x_\alpha^\beta}\right)\right\}_{\beta=1}^{\Lambda<\infty}$ of $V$. For simplicity, by an abuse of notation, we remove the excessive indices and rewrite the finite subcover of $V$ as $\{(\rho_\alpha, W_\alpha)\}_{\alpha=1}^{\Lambda} := \left\{\rho_\alpha\left(W_{x_\alpha^\beta}\right)\right\}_{\beta=1}^{\Lambda}$.

For each $\alpha \in [1, \ldots, \Lambda]$, let $S_\alpha := \{\nu_\alpha := \nu_{\theta^\alpha} | \theta^\alpha \in \Xi_\alpha \subset \mathbb{R}^{m_\alpha}\} \subset \text{Prob}(W_\alpha)$ denote a family of finitely parametrized volume form over $W_\alpha$ parametrized by $\theta^\alpha \in \Xi_\alpha \subset \mathbb{R}^{m_\alpha}$, $m_\alpha \in \mathbb{N}_+$. Note that in general we allow the parametric families $S_\alpha$ to have different parametrizations and dimensions $m_\alpha$. Furthermore, consider for each $\alpha$ the induced family of local probability densities $\tilde{S}_\alpha := \rho_\alpha^{-1^*} S_\alpha = \left\{\tilde{\nu}_\alpha = \rho_\alpha^{-1^*} \nu_\alpha\right\} \subset \text{Prob}(M) \subset \text{Vol}(M)$ on $\rho_\alpha(W_\alpha) \subset M$ (see Sect. 5.2.1). We define **parametrized mixture densities over** $V \subset M$ by patching together the locally induced ones as follows:

$$\tilde{\nu} := \sum_{\alpha=1}^{\Lambda} \varphi_\alpha \cdot \tilde{\nu}_\alpha, \tag{6.2}$$

where $\varphi_\alpha \in [0, 1]$ for all $\alpha \in [1, \dots, \Lambda]$, and $\sum_{\alpha=1}^{\Lambda} \varphi_\alpha = 1$. Let $\overline{S}_0$ denote the **closure of the simplex of mixture coefficients**:

$$\overline{S}_0 := \left\{ \{\varphi_\alpha\}_{\alpha=1}^{\Lambda} \,|\varphi_\alpha \in [0, 1], \sum_{\alpha=1}^{\Lambda} \varphi_\alpha = 1 \right\} \subset [0, 1]^{\Lambda-1}$$

**Remark 6.2** The closure of simplex of mixture coefficients $\overline{S}_0$ has the structure of a dually flat statistical manifold using the modified Fisher metric (to be defined in Definition 8.1) and Bregman divergence pair discussed in Sect. 8.3. We denote the set of parameters of $\overline{S}_0$ by $\Xi_0 := \left\{\theta^0\right\}$.

The **set of mixture densities** $\mathcal{L}_V$ is denoted by:

$$\mathcal{L}_V := \left\{ \tilde{\nu} = \sum_{\alpha=1}^{\Lambda} \varphi_\alpha \cdot \tilde{\nu}_\alpha \mid \{\varphi_\alpha\}_{\alpha=1}^{\Lambda} \in \overline{S}_0, \; \tilde{\nu}_\alpha \in \tilde{S}_\alpha \right\}.$$

For $x \in M$, the mixture volume form $\tilde{\nu}$ can be expressed in local coordinates $\left(x^1, \dots, x^n\right)$ around $x \in M$ as:

$$\tilde{\nu}|_x = \tilde{p}(x, \xi) dx^1 \wedge \dots \wedge dx^n = \sum_{\alpha=1}^{\Lambda} \varphi_\alpha \cdot \tilde{p}_\alpha(x, \theta^\alpha) dx^1 \wedge \dots \wedge dx^n, \tag{6.3}$$

where $\tilde{p}_\alpha(x, \theta^\alpha)$ are parametrized probability density functions naturally associated to $\tilde{\nu}_\alpha$, parametrized by sets of parameters $\theta^\alpha := \left(\theta_i^\alpha\right)_{i=1}^{m_\alpha} \in \Xi_\alpha$ (see Remark 5.4). In local coordinates $\left(x^1, \dots, x^n\right)$ of $M$, $\tilde{p}_\alpha(x, \theta^\alpha)$ is defined implicitly by $\tilde{p}_\alpha(x, \theta^\alpha) dx^1 \wedge \dots \wedge dx^n := \tilde{\nu}_\alpha$.[1] The parameters of the mixture distribution $\tilde{p}(x, \xi)$ are collected in $(\xi_i)_{i=1}^{d} := \left(\theta^0, \theta^1, \dots, \theta^\Lambda\right) \in \Xi_0 \times \Xi_1 \times \dots \times \Xi_\Lambda$, where $d := \dim\left(\Xi_0 \times \Xi_1 \times \dots \times \Xi_\Lambda\right)$ denote the dimension of $\mathcal{L}_V$.

**Remark 6.3** (Exhaustion by Compact Set and the Extent of Mixture Densities) Even though at first glance totally bound-ness of $V \subset M$ might be quite restrictive,

[1] When $M$ is a Riemannian manifold, we may choose to use the Riemannian volume form $dV_g$ instead of $dx^1 \wedge \dots \wedge dx^n$. In which case we obtain $\tilde{p}_\alpha(x, \theta^\alpha)$ by: $\tilde{p}_\alpha(x, \theta^\alpha) dV_g := \tilde{\nu}_\alpha$.

we show that it allows us to approximate the manifold sufficiently well. Since every smooth topological manifold is $\sigma$-locally compact, $M$ admits a compact exhaustion.

**Definition 6.2** An **exhaustion by compact sets** is an increasing sequence of compact subsets $\overline{W}^k$ of $M$ such that $\phi \neq \overline{W}^1 \subset \text{int}(\overline{W}^2) \subset \overline{W}^2 \subseteq \cdots \subseteq M$ and:

$$\lim_{k\to\infty} \overline{W}^k = \bigcup_{k=1}^{\infty} \overline{W}^k = M.$$

In particular, for any totally bounded subset $V \subsetneq M$, there exists $K \in \mathbb{N}$ such that for all $k > K$:

$$\overline{W^k}\ (\supset W^k) \supset V.$$

Therefore we may extend totally bounded subsets, the support of the mixture distributions, arbitrary close to the entire manifold $M$.

## 6.3 Geometrical Structure of $\mathcal{L}_V$

In this section we detail the geometrical structure of family of mixture densities $\mathcal{L}_V$ defined over totally bounded subsets $V$ of $M$. We first show that $\mathcal{L}_V$ is a smooth manifold under two natural conditions. We then show that $\mathcal{L}_V$ is a product Riemannian manifold of the closure of simplex of mixture cofficients and the locally inherited families of component densities. The product dualistic Riemannian structure on $\mathcal{L}_V$ is given by a "mixture divergence" on $\mathcal{L}_V$ constructed by the divergence functions on the simplex of mixture coefficients and families of mixture component densities. Furthermore, if the families of mixture component densities are all dually flat, we show that $\mathcal{L}_V$ is also dually flat with the canonical divergence is given by the mixture divergence.

Consider a totally bounded subset $V \subset M$ and a finite orientation-preserving open cover $\{(\rho_\alpha, W_\alpha)\}_{\alpha=1}^{\Lambda}$ of $V$ by metric balls.

For each $\alpha = 1, \ldots, \Lambda$, let $\left(S_\alpha, g_\alpha, \nabla^\alpha, \nabla^{\alpha^*}\right) \subset \text{Prob}(W_\alpha)$ be a family of parametrized probability densities over $W_\alpha$ with the corresponding statistical manifold structure. Let $\tilde{S}_\alpha := \rho_\alpha^{-1^*} S_\alpha$ denote the pulled-back family of probability densities on $\rho_\alpha(W_\alpha) \subset M$ with corresponding pulled-back torsion-free dualistic structures $\left(\tilde{g}_\alpha, \tilde{\nabla}^\alpha, \tilde{\nabla}^{\alpha^*}\right)$ (see Sects. 5.1 and 5.2.1). Let $\overline{S}_0$ denote the closure of simplex of mixture coefficients with corresponding *dually flat* torsion-free dualistic structure $\left(g_0, \nabla^0, \nabla^{0^*}\right)$ (see for example the dually flat dualistic structure defined by the Bregman divergence defined in Eq. (8.20)).

### 6.3.1 $\mathcal{L}_V$ as a Smooth Manifold

In this section we show that the family of mixture densities $\mathcal{L}_V$ over totally bounded subsets $V$ of Riemannian manifolds $M$ is itself a smooth manifold under two natural conditions. In particular, we show that the parametrization map of $\mathcal{L}_V$ is an injective immersion, which makes $\mathcal{L}_V$ an immersion submanifold of Prob$(M)$.

For simplicity, let $\tilde{p}_\alpha(x) := \tilde{p}(x, \theta^\alpha)$ denote the probability density function corresponding to $\tilde{\nu}_\alpha \in \tilde{S}_\alpha$ for all $\alpha = 1, \ldots, \Lambda$. We show that $\mathcal{L}_V$ is a smooth manifold under the following two natural conditions:

1. **Family of mixture component distributions have different proper support**: Let $V_\alpha := \left\{x \in M \mid \tilde{p}_\alpha(x) > 0, \ \forall \tilde{p}_\alpha \in \tilde{S}_\alpha\right\}$ denote the proper support of probability densities $\tilde{p}_\alpha \in \tilde{S}_\alpha$ for each $\alpha \in \{1, \ldots, \Lambda\}$. We assume $V_\alpha \setminus V_\beta \neq \emptyset$ for $\beta \neq \alpha$.
2. **No functional dependency between mixture component densities**: We construct mixture densities in $\mathcal{L}_V$ as unconstrained mixtures, meaning there is no functional dependency between mixture component. In other words, changing parameters $\theta^\beta \in \Xi_\beta$ of mixture component $\tilde{p}_\beta \in \tilde{S}_\beta$ has no influence on $\tilde{p}_\alpha \in \tilde{S}_\alpha$ for $\beta \neq \alpha$ and vice versa. Formally, we write this condition as follows: For each $\tilde{p}_\alpha \in \tilde{S}_\alpha$, $\frac{\partial \tilde{p}_\alpha}{\partial \theta_k^\beta} = 0, \quad \forall \theta_k^\beta \in \Xi^\beta \subset \mathbb{R}^{m_\beta}, \forall \beta \neq \alpha$.

**Remark 6.4** 1. Condition 1 can always be satisfied simply by choosing a suitable open cover of $V$.
2. Condition 2 is automatically fulfilled for unconstrained mixture models. One can imagine introducing functional dependencies among mixture component distributions, but this is not the case considered here. We make the assumption that: if we alter one distribution $\tilde{p}_\alpha \in \tilde{S}_\alpha$, it does not affect distributions in $\tilde{S}_\beta$ for $\beta \neq \alpha$.

We now discuss the implications of Conditions 1 and 2 in further detail:

Condition 1 implies the component distributions $\tilde{p}_\alpha \in \tilde{S}_\alpha$ are linearly independent functions, and the local parametrization map of the $\overline{S}_0$, denoted by $\theta^0 \in \Xi_0 \mapsto \{\varphi_\alpha\}_{\alpha=1}^\Lambda \mapsto \sum_{\alpha=1}^\Lambda \varphi_\alpha \tilde{p}_\alpha(x, \theta^\alpha)$, is injective.[2]

Condition 2 implies the following: Consider two distributions $\tilde{p}, \tilde{q} \in \mathcal{L}_V$ sharing the same mixture coefficients $\{\varphi_\alpha\}_{\alpha=1}^\Lambda$, i.e. $\tilde{p}(x) = \sum_{\alpha=1}^\Lambda \varphi_\alpha \tilde{p}_\alpha(x)$ and $\tilde{q} = \sum_{\alpha=1}^\Lambda \varphi_\alpha \tilde{q}_\alpha(x)$. If $\tilde{p}(x) = \tilde{q}(x)$ for all $x \in M$:

$$\frac{\partial \tilde{p}(x)}{\partial \theta_i^\alpha} = \frac{\partial \tilde{q}(x)}{\partial \theta_i^\alpha}, \quad \forall \alpha, \forall i$$
$$\frac{\partial^\ell \tilde{p}(x)}{\left(\partial \theta_i^\alpha\right)^\ell} = \frac{\partial^\ell \tilde{q}(x)}{\left(\partial \theta_i^\alpha\right)^\ell}, \quad \forall \ell \in \mathbb{N}_+,$$

hence by Condition 2, for each $\alpha \in [1, \ldots, \Lambda]$, there exists a constant $c_\alpha$ such that:

[2] Note that the mixture component densities $\tilde{p}_\alpha(x, \theta^\alpha)$ are fixed.

$$\varphi_\alpha \tilde{p}_\alpha(x) = \varphi_\alpha \tilde{q}_\alpha(x) + c_\alpha, \quad \forall x \in M.$$

Since $\tilde{p}_\alpha$ and $\tilde{q}_\alpha$ are probability densities, they integrate to 1 over $M$, and we have the following:

$$\varphi_\alpha \cdot \underbrace{\int_M \tilde{p}_\alpha}_{=1} = \varphi_\alpha \cdot \underbrace{\int_M \tilde{q}_\alpha}_{=1} + \int_M c_\alpha.$$

This means $\int_M c_\alpha = c_\alpha \cdot \int_M 1 = 0$. Since $M$ is orientable we have $\int_M 1 \neq 0$, which implies $c_\alpha = 0$ and $\tilde{p}_\alpha = \tilde{q}_\alpha$ for all $\alpha = 1, \ldots, \Lambda$.

Hence by injectivity of the parametrization mapping of the local mixture component families $\theta^\alpha \mapsto \tilde{p}_\alpha$ of $\tilde{S}_\alpha$ for $\alpha \in [1, \ldots, \Lambda]$, the parametrization of mixture component parameters $(\theta^1, \ldots, \theta^\Lambda) \mapsto (\tilde{p}_1, \ldots, \tilde{p}_\Lambda)$ is injective as well. This implies the local parametrization map $\xi : \left(\theta^0, \theta^1, \ldots, \theta^\Lambda\right) \in \Xi_0 \times \Xi_1 \times \cdots \times \Xi_\Lambda \mapsto \tilde{p}(x) = \sum_{\alpha=1}^{\Lambda} \varphi_\alpha \tilde{p}_\alpha(x)$ onto $\mathcal{L}_V$ is also injective.

We now show that the local parametrization map $\xi : \Xi_0 \times \Xi_1 \times \cdots \times \Xi_\Lambda \mapsto \mathcal{L}_V$ is an immersion. By Condition 2, the parameters are also independent in the sense that for $\beta \neq \alpha$:

$$0 = \frac{\partial \tilde{p}_\alpha}{\partial \theta_k^\beta} = \frac{\partial \tilde{p}_\alpha}{\partial \theta_i^\alpha} \cdot \frac{\partial \theta_i^\alpha}{\partial \theta_k^\beta} \quad \forall \tilde{p}_\alpha \in \tilde{S}_\alpha \Leftrightarrow \frac{\partial \theta_i^\alpha}{\partial \theta_k^\beta} = 0.$$

Moreover, let $\left(\overline{\theta}_i^\alpha := \theta_i^\alpha \circ \left(\rho^{-1^*}\right)^{-1} = \theta_i^\alpha \circ \rho^*\right)$ on $\tilde{S}_\alpha$ denote the pulled-back local coordinates maps (see Sect. 5.2.1), then:

$$\frac{\partial \theta_i^\alpha \circ \rho^*}{\partial \theta_k^\beta \circ \rho^*} = \left(\rho^{-1^*}\right)_* \frac{\partial}{\partial \theta_k^\beta}\left(\theta_i^\alpha \circ \rho^*\right) = \frac{\partial}{\partial \theta_k^\beta}\left(\theta_i^\alpha \circ \rho^* \circ \rho^{-1^*}\right) = 0.$$

In other words, pullback by $\rho$ does not introduce additional functional dependencies among parameters.

Let $\xi_0 : \theta^0 \in \Xi_0 \mapsto \{\varphi_\alpha\}_{\alpha=1}^{\Lambda} \mapsto \sum_{\alpha=1}^{\Lambda} \varphi_\alpha \tilde{p}_\alpha$ denote the parametrization of $\overline{S}_0$ and let $\xi_\alpha : \tilde{\theta}^\alpha \in \Xi_\alpha \mapsto \tilde{S}_\alpha$ denote the parametrization map of $\tilde{S}_\alpha$ for $\alpha = 1, \ldots \Lambda$. The Jacobian matrix of the parametrization map $\xi : (\theta^0, \theta^1, \ldots, \theta^\Lambda) \in \Xi_0 \times \Xi_1 \times \cdots \Xi_\Lambda \mapsto \sum_{\alpha=1}^{\Lambda} \varphi_\alpha \tilde{p}_\alpha(x, \theta^\alpha) =: \tilde{p}$ is thus given by:

$$\mathrm{Jac}_\xi(\tilde{p}) = \left[\begin{array}{c|cccccc} \mathrm{Jac}_{\xi_0}(\tilde{p}) & & & & 0 & & \\ \hline & \mathrm{Jac}_{\xi_1}(\tilde{p}) & 0 & \cdots & \cdots & 0 \\ & 0 & \mathrm{Jac}_{\xi_2}(\tilde{p}) & 0 & \cdots & 0 \\ 0 & \vdots & \ddots & \ddots & \ddots & 0 \\ & 0 & \cdots & 0 & \mathrm{Jac}_{\xi_{\Lambda-1}}(\tilde{p}) & 0 \\ & 0 & 0 & \cdots & 0 & \mathrm{Jac}_{\xi_\Lambda}(\tilde{p}) \end{array}\right].$$

$\overline{S}_0$ can be endowed with a statistical manifold structure (Remark 6.2),[3] and by the discussion of Sect. 5.2, $\tilde{S}_\alpha$ are statistical manifolds for $\alpha = 1, \ldots, \Lambda$. Since parametrization maps of statistical manifolds are immersions [CU14], the submatrices $\mathrm{Jac}_{\xi_0}(\tilde{p}), \mathrm{Jac}_{\xi_1}(\tilde{p}), \ldots, \mathrm{Jac}_{\xi_\Lambda}(\tilde{p})$ are injective, and the Jacobian matrix $\mathrm{Jac}_\xi(\tilde{p})$ described above is injective as well. This means the local parametrization map $\xi : \Xi_0 \times \Xi_1 \times \cdots \times \Xi_\Lambda \mapsto \mathcal{L}_V$ is an immersion.

The set of mixture densities $\mathcal{L}_V$ over totally bounded subsets $V$ of $M$ is therefore an *immersion submanifold* of $\mathrm{Prob}(M)$, where the mixture densities $\tilde{p} \in \mathcal{L}_V$ are identified by the local coordinates $\mathcal{L}_V \ni \tilde{p} := \sum_{\alpha=1}^{\Lambda} \varphi_\alpha \tilde{p}_\alpha(x, \theta^\alpha) \mapsto (\theta^0, \theta^1, \ldots, \theta^\Lambda) \in \Xi_0 \times \Xi_1 \times \cdots \Xi_\Lambda$.

### 6.3.2 *Torsion-Free Dualistic Structure on $\mathcal{L}_V$*

In this section we construct a torsion-free dualistic structure on $\mathcal{L}_V$ by considering the following **mixture divergence** function on $\mathcal{L}_V \times \mathcal{L}_V$:

$$D : \mathcal{L}_V \times \mathcal{L}_V \to \mathbb{R}$$

$$D(\tilde{p}, \tilde{q}) = D\left(\sum_{\alpha=1}^{\Lambda} \varphi_\alpha \tilde{p}_\alpha, \sum_{\alpha=1}^{\Lambda} \varphi'_\alpha \tilde{q}_\alpha\right) := D_0\left(\{\varphi_\alpha\}_{\alpha=1}^{\Lambda}, \{\varphi'_\alpha\}_{\alpha=1}^{\Lambda}\right) + \sum_{\alpha=1}^{\Lambda} \overline{D}_\alpha(\tilde{p}_\alpha, \tilde{q}_\alpha), \tag{6.4}$$

where $D_0$ is a divergence $\overline{S}_0$ (see for example the Bregman divergence defined in Eq. (8.20)), and $\overline{D}_\alpha$ is the induced divergence on smooth manifolds $\tilde{S}_\alpha$ described by Remark 5.2 in Sect. 5.1. It is immediate by construction that $D$ satisfies the conditions of a divergence function (Definition 5.1):

1. **Non-negativity**: Since $\overline{D}_\alpha$'s and $D_0$ are both non-negative, so is $D$:

$$D = \underbrace{D_0}_{\geq 0} + \sum_{\alpha=1}^{\Lambda} \underbrace{\overline{D}_\alpha}_{\geq 0}$$

2. **Identity**: Since $\overline{D}_\alpha$'s and $D_0$ are divergences, the following is satisfied:

$$D = 0 \Leftrightarrow \begin{Bmatrix} D_0(\varphi) = 0 \\ \overline{D}_\alpha(\tilde{p}_\alpha, \tilde{q}_\alpha) = 0 \quad \forall \alpha \end{Bmatrix} \Leftrightarrow \begin{Bmatrix} \varphi_\alpha = \varphi'_\alpha \\ \tilde{p}_\alpha = \tilde{q}_\alpha \end{Bmatrix} \forall \alpha \Leftrightarrow \tilde{p} = \tilde{q}.$$

Given a divergence function on a manifold, one can construct the corresponding (torsion-free) dualistic Riemannian structure on the manifold [E+92, AN00]. In par-

[3] This is further described in Sect. 8.3.

ticular, the torsion-free dualistic structure on $\mathcal{L}_V$ induced by the mixture divergence $D$ (Eq. (6.4)) is derived in the following result.

**Theorem 6.1** *The torsion-free dualistic structure $\left\{g^D, \nabla^D, \nabla^{D^*}\right\}$ of $\mathcal{L}_V$ generated by the mixture divergence $D$ (desribed by Eq. (6.4)), is given by:*

$$\left\{g^{D_0} \oplus \bigoplus_{\alpha=1}^{\Lambda} g^{\overline{D}_\alpha}, \nabla^{D_0} \oplus \bigoplus_{\alpha=1}^{\Lambda} \nabla^{\overline{D}_\alpha}, \nabla^{D_0^*} \oplus \bigoplus_{\alpha=1}^{\Lambda} \nabla^{\overline{D}_\alpha^*}\right\}.$$

***Proof*** For $\alpha \in [0, 1, \ldots, \Lambda]$, let $\theta^\alpha \in \Xi^\alpha \subset \mathbb{R}^{m_\alpha}$ denote the local coordinates of $S_\alpha$. Let $(\xi_i)_{i=1}^d := (\theta^0, \tilde{\theta}^1, \ldots, \tilde{\theta}^\Lambda)$ denote the coordinates of $\mathcal{L}_V$, where $d := \dim(\Xi_0 \times \Xi_1 \times \cdots \times \Xi_\Lambda)$ and $\left(\tilde{\theta}_j^\alpha := \theta_j^\alpha \circ \rho_\alpha^*\right)_{j=1}^{m_\alpha}$ denote the local pulled-back coordinates of $\tilde{S}_\alpha$ discussed in Sect. 5.1.1 respectively (see also Remark 5.3 and 5.4). We write the local coordinate frame corresponding to $(\xi_i)_{i=1}^d$ as $\left(\overline{\partial}_i := \frac{\partial}{\partial \xi_i}\right)$. We can then construct a dualistic structure $\left(g^D, \nabla^D, \nabla^{D^*}\right)$ on $\mathcal{L}_V$ corresponding to the mixture divergence function $D$ (defined in Eq. (6.4)) as follows. First of all, the matrix $\left[g_{ij}^D\right]$ corresponding to the induced Riemannian metric $g^D$ is given by:

$$\begin{aligned}
g_{ij}^D(\tilde{p}) := g^D\left(\overline{\partial}_i, \overline{\partial}_j\right)\Big|_{\tilde{p}} &= -D[\overline{\partial}_i; \overline{\partial}_j]\big|_{\tilde{p}} \\
&= -D_0[\overline{\partial}_i; \overline{\partial}_j]\big|_{\tilde{p}} + \sum_{\alpha=1}^{\Lambda} -\overline{D}_\alpha[\overline{\partial}_i; \overline{\partial}_j]\Big|_{\tilde{p}} \\
&= g^{D_0}\left(\overline{\partial}_i, \overline{\partial}_j\right)\Big|_{\tilde{p}} + \sum_{\alpha=1}^{\Lambda} g^{\overline{D}_\alpha}\left(\overline{\partial}_i, \overline{\partial}_j\right)\Big|_{\tilde{p}}. \qquad (6.5)
\end{aligned}$$

Equation (6.5) can thus be computed in four cases:

1. If both the input of Eq. (6.5) $\left(\overline{\partial}_i, \overline{\partial}_j\right)$ belong to $\mathcal{E}\left(T\overline{S}_0\right)$: i.e. if $\overline{\partial}_i = \frac{\partial}{\partial \theta_i^0}$ and $\overline{\partial}_j = \frac{\partial}{\partial \theta_j^0}$ then for $\{\varphi_\alpha\}_{\alpha=1}^\Lambda \in \overline{S}_0$:

$$\begin{aligned}
g^D\left(\overline{\partial}_i, \overline{\partial}_j\right)\Big|_{\tilde{p}} &= g^{D_0}\left(\overline{\partial}_i, \overline{\partial}_j\right)\Big|_{\tilde{p}} + 0 \\
&= g^{D_0}\left(\overline{\partial}_i, \overline{\partial}_j\right)\Big|_{\{\varphi_\alpha\}_{\alpha=1}^\Lambda}. \qquad (6.6)
\end{aligned}$$

2. If both the input of Eq. (6.5) $\left(\overline{\partial}_i, \overline{\partial}_j\right)$ belong to the *same* $\mathcal{E}\left(T\tilde{S}_\beta\right)$: i.e. If $\overline{\partial}_i = \tilde{\partial}_i^\beta = \frac{\partial}{\partial \tilde{\theta}_i^\beta}$ and $\overline{\partial}_j = \tilde{\partial}_j^\beta = \frac{\partial}{\partial \tilde{\theta}_j^\beta}$, where $\left(\tilde{\theta}_j^\beta\right)_{j=1}^{m_\beta}$ denote the local parametrization of $\tilde{S}_\beta$ for some $\beta \in \{1, \ldots, \Lambda\}$. Then by Condition 2:

$$
\begin{aligned}
g^D\left(\overline{\partial}_i, \overline{\partial}_j\right)\big|_{\tilde{p}} &= 0 + \sum_{\alpha=1}^{\Lambda} g^{\overline{D}_\alpha}\left(\tilde{\partial}_i^\beta, \tilde{\partial}_j^\beta\right)\Big|_{\tilde{p}_\alpha} \\
&= 0 + \sum_{\alpha=\beta} g^{\overline{D}_\alpha}\left(\tilde{\partial}_i^\beta, \tilde{\partial}_j^\beta\right)\Big|_{\tilde{p}_\alpha} \\
&= g^{\overline{D}_\beta}\left(\tilde{\partial}_i^\beta, \tilde{\partial}_j^\beta\right)\Big|_{\tilde{p}_\beta} = g^\beta(\partial_i^\beta, \partial_j^\beta)\Big|_{p_\beta},
\end{aligned} \tag{6.7}
$$

where $g^\beta$ is the metric on $S_\beta$, and $p_\beta := \left(\rho^{-1^*}\right)^{-1} \tilde{p}_\beta = \rho^* \tilde{p}_\beta$. The last equality is due to Theorem 5.3.

3. If the first input of Eq. (6.5) belongs to $\mathcal{E}\left(T\overline{S}_0\right)$ and the second input belongs to $\mathcal{E}\left(T\tilde{S}_\beta\right)$ (or vice versa by symmetry): i.e. If $\overline{\partial}_i = \frac{\partial}{\partial\theta_i^0}$ and $\overline{\partial}_j = \tilde{\partial}_j^\beta = \frac{\partial}{\partial\tilde{\theta}_j^\beta}$ where $\left(\tilde{\theta}_j^\beta\right)_{j=1}^{m_\beta}$ denote the local parametrization of $\tilde{S}_\beta$ for some $\beta \in \{1, \ldots, \Lambda\}$, then

$$
\begin{aligned}
g^D\left(\overline{\partial}_i, \overline{\partial}_j\right)\big|_{\tilde{p}} &= g^{D_0}\left(\overline{\partial}_i, \pi_{*0}\tilde{\partial}_j^\beta\right)\Big|_{\{\varphi_\alpha\}_{\alpha=1}^{\Lambda}} + g^{\overline{D}_\beta}\left(\pi_{*_\beta}\overline{\partial}_i, \tilde{\partial}_j^\beta\right)\Big|_{\tilde{p}_\beta} \\
&= g^{D_0}\left(\overline{\partial}_i, 0\right)\big|_{\{\varphi_\alpha\}_{\alpha=1}^{\Lambda}} + g^{\overline{D}_\beta}\left(0, \tilde{\partial}_j^\beta\right)\Big|_{\tilde{p}_\beta} = 0 + 0 = 0,
\end{aligned}
$$

where $\pi_{*_0} : \mathcal{E}\left(T\mathcal{L}_V\right) \to \mathcal{E}\left(T\overline{S}_0\right)$ and $\pi_{*_\beta} : \mathcal{E}\left(T\mathcal{L}_V\right) \to \mathcal{E}\left(T\tilde{S}_\beta\right)$ denote the natural projections onto the corresponding tangent subbundles.

4. Otherwise if the first and second input of Eq. (6.5) belongs to *different* $\mathcal{E}\left(T\tilde{S}_\alpha\right)$ and $\mathcal{E}\left(T\tilde{S}_\beta\right)$: i.e. If $\overline{\partial}_i = \tilde{\partial}_i^\alpha = \frac{\partial}{\partial\tilde{\theta}_i^\alpha}$ and $\overline{\partial}_j = \tilde{\partial}_j^\beta = \frac{\partial}{\partial\tilde{\theta}_j^\beta}$, where $\left(\tilde{\theta}_j^\alpha\right)_{j=1}^{m_\alpha}$ and $\left(\tilde{\theta}_j^\beta\right)_{j=1}^{m_\beta}$ denote the local parametrization of $\tilde{S}_\alpha$ and $\tilde{S}_\beta$ respectively for some $\alpha \neq \beta \in \{1, \ldots, \Lambda\}$. By a similar argument of the previous case, Eq. (6.5) then becomes:

$$
\begin{aligned}
g^D\left(\overline{\partial}_i, \overline{\partial}_j\right)\big|_{\tilde{p}} &= 0 + g^{\overline{D}_\alpha}\left(\tilde{\partial}_i^\alpha, \pi_{*_\alpha}\tilde{\partial}_j^\beta\right)\Big|_{\tilde{p}_\alpha} + g^{\overline{D}_\beta}\left(\pi_{*_\beta}\tilde{\partial}_i^\alpha, \tilde{\partial}_j^\beta\right)\Big|_{\tilde{p}_\beta} \\
&= 0 + g^{\overline{D}_\alpha}\left(\tilde{\partial}_i^\alpha, 0\right)\Big|_{\tilde{p}_\alpha} + g^{\overline{D}_\beta}\left(0, \tilde{\partial}_j^\beta\right)\Big|_{\tilde{p}_\beta} = 0 + 0 + 0 = 0.
\end{aligned}
$$

The Christoffel symbols of the connection $\nabla^D$ induced by $D$ are given by:

$$
\begin{aligned}
\Gamma_{ijk}^D\Big|_{\tilde{p}} := \langle\nabla_{\overline{\partial}_i}^D\overline{\partial}_j, \overline{\partial}_k\rangle_{g^D}\Big|_{\tilde{p}} &= -D\left[\overline{\partial}_i\overline{\partial}_j; \overline{\partial}_k\right]\big|_{\tilde{p}} \\
&= -D_0[\overline{\partial}_i\overline{\partial}_j; \overline{\partial}_k]\big|_{\tilde{p}} + \sum_{\alpha=1}^{\Lambda} -\overline{D}_\alpha[\overline{\partial}_i\overline{\partial}_j; \overline{\partial}_k]\big|_{\tilde{p}} \\
&= \langle\nabla_{\overline{\partial}_i}^{D_0}\overline{\partial}_j, \overline{\partial}_k\rangle_{g^D}\Big|_{\tilde{p}} + \sum_{\alpha=1}^{\Lambda} \langle\nabla_{\overline{\partial}_i}^{\overline{D}_\alpha}\overline{\partial}_j, \overline{\partial}_k\rangle_{g^D}\Big|_{\tilde{p}},
\end{aligned} \tag{6.8}
$$

where $-D[\partial_i\partial_j;\partial_k] := -\partial_i^1\partial_j^1\partial_k^2\, D[p;q]|_{q=p}$ (by Eq. (5.3) in Sect. 5.1). By the similar four case argument in the derivation of the induced metric $g^D$ above, we obtain the following[4]:

$$\langle\nabla^D_{\overline{\partial}_i}\overline{\partial}_j,\overline{\partial}_k\rangle_{g^D}\Big|_{\tilde{p}} = \begin{cases} \langle\nabla^{D_0}_{\overline{\partial}_i}\overline{\partial}_j,\overline{\partial}_k\rangle_{g^{D_0}}\Big|_{\{\varphi_\alpha\}_{\alpha=1}^{\Lambda}} & \text{if } \overline{\partial}_\ell = \frac{\partial}{\partial\theta_\ell^0},\ \text{for } \ell = i,j,k\ , \\ \langle\nabla^{\overline{D}_\beta}_{\overline{\partial}_i}\overline{\partial}_j,\overline{\partial}_k\rangle_{g^{\overline{D}_\beta}}\Big|_{\tilde{p}_\beta} & \text{if } \overline{\partial}_\ell = \tilde{\partial}_\ell^\beta = \frac{\partial}{\partial\tilde{\theta}_\ell^\beta},\ \text{for } \begin{matrix}\ell = i,j,k \\ \beta = 1,\ldots,\Lambda\end{matrix}, \\ 0 & \text{Otherwise.} \end{cases} \tag{6.9}$$

For $\alpha \in [1,\ldots,\Lambda]$, the above results imply that for vector fields $X_0, Y_0 \in \mathcal{E}\left(T\overline{S}_0\right)$, $X_\alpha, Y_\alpha \in \mathcal{E}\left(T\tilde{S}_\alpha\right)$, we have the following:

$$\begin{aligned}\nabla^D_{\left(Y_0+\sum_{\alpha=1}^{\Lambda}Y_\alpha\right)}\left(X_0+\sum_{\alpha=1}^{\Lambda}X_\alpha\right) &= \nabla^D\left(X_0+\sum_{\alpha=1}^{\Lambda}X_\alpha, Y_0+\sum_{\alpha=1}^{\Lambda}Y_\alpha\right) \\ &= \nabla^{D_0}_{Y_0}X_0 + \sum_{\alpha=1}^{\Lambda}\nabla^{\overline{D}_\alpha}_{Y_\alpha}X_\alpha. \end{aligned} \tag{6.10}$$

By symmetry and the fact that $g^D = g^{D^*}$ [AN00], we have the following for the induced dual connection $\nabla^{D^*}$:

$$\langle\nabla^{D^*}_{\overline{\partial}_i}\overline{\partial}_j,\overline{\partial}_k\rangle_{g^D}\Big|_{\tilde{p}} = \langle\nabla^{D_0^*}_{\overline{\partial}_i}\overline{\partial}_j,\overline{\partial}_k\rangle_{g^D}\Big|_{\tilde{p}} + \sum_{\alpha=1}^{\Lambda}\langle\nabla^{\overline{D}_\alpha^*}_{\overline{\partial}_i}\overline{\partial}_j,\overline{\partial}_k\rangle_{g^D}\Big|_{\tilde{p}}. \tag{6.11}$$

Furthermore, we obtain the following result analogous to Eq. (6.9):

$$\langle\nabla^{D^*}_{\overline{\partial}_i}\overline{\partial}_j,\overline{\partial}_k\rangle_{g^D}\Big|_{\tilde{p}} = \begin{cases} \langle\nabla^{D_0^*}_{\overline{\partial}_i}\overline{\partial}_j,\overline{\partial}_k\rangle_{g^{D_0}}\Big|_{\{\varphi_\alpha\}_{\alpha=1}^{\Lambda}} & \text{if } \overline{\partial}_\ell = \frac{\partial}{\partial\theta_\ell^0},\ \text{for } \ell = i,j,k \\ \langle\nabla^{\overline{D}_\beta^*}_{\overline{\partial}_i}\overline{\partial}_j,\overline{\partial}_k\rangle_{g^{\overline{D}_\beta}}\Big|_{\tilde{p}_\beta} & \text{if } \overline{\partial}_\ell = \tilde{\partial}_\ell^\beta = \frac{\partial}{\partial\tilde{\theta}_\ell^\beta},\ \text{for } \begin{matrix}\ell = i,j,k \\ \beta = 1,\ldots,\Lambda\end{matrix}, \\ 0 & \text{Otherwise.} \end{cases} \tag{6.12}$$

Finally, for $X_0, Y_0 \in \mathcal{E}\left(T\overline{S}_0\right)$, $X_\alpha, Y_\alpha \in \mathcal{E}\left(T\tilde{S}_\alpha\right)$ for $\alpha \in [1,\ldots,\Lambda]$, we obtain:

$$\begin{aligned}\nabla^{D^*}_{\left(Y_0+\sum_{\alpha=1}^{\Lambda}Y_\alpha\right)}\left(X_0+\sum_{\alpha=1}^{\Lambda}X_\alpha\right) &= \nabla^{D^*}\left(X_0+\sum_{\alpha=1}^{\Lambda}X_\alpha, Y_0+\sum_{\alpha=1}^{\Lambda}Y_\alpha\right) \\ &= \nabla^{D_0^*}_{Y_0}X_0 + \sum_{\alpha=1}^{\Lambda}\nabla^{\overline{D}_\alpha^*}_{Y_\alpha}X_\alpha. \end{aligned} \tag{6.13}$$

[4] The third line of Eq. (6.9) encapsulates the last two cases of $g^D$.

Therefore by Eqs. (6.5), (6.10), (6.13), the dualistic structure $\{g^D, \nabla^D, \nabla^{D^*}\}$ induced by $D$ naturally decomposes into the parts corresponding to the mixture coefficients and mixture components. We abbreviate Eqs. (6.5), (6.10), (6.13) by the following compact form:

$$\{g^D, \nabla^D, \nabla^{D^*}\} = \left\{ g^{D_0} \oplus \bigoplus_{\alpha=1}^{\Lambda} g^{\overline{D}_\alpha}, \nabla^{D_0} \oplus \bigoplus_{\alpha=1}^{\Lambda} \nabla^{\overline{D}_\alpha}, \nabla^{D_0^*} \oplus \bigoplus_{\alpha=1}^{\Lambda} \nabla^{\overline{D}_\alpha^*} \right\}. \quad (6.14)$$

Finally, since the dualistic structures $(g_0, \nabla^0, \nabla^{0^*})$ and $\left(\tilde{g}_\alpha, \tilde{\nabla}^\alpha, \tilde{\nabla}^{\alpha^*}\right)$ on $\overline{S}_0$ and $\tilde{S}_\alpha$ respectively for $\alpha \in [1, \ldots, \Lambda]$ are torsion-free, so is $\{g^D, \nabla^D, \nabla^{D^*}\}$. □

## 6.4 $\mathcal{L}_V$ as a Product Statistical Manifold

In this section we show that $\mathcal{L}_V$ is a product Riemannian manifold. We recall properties of product Riemannian manifolds from the literature in [Sak96, dC92, Lee06].

**Remark 6.5** Since $\mathcal{L}_V$ consists of a finite mixture of probability distributions, to show that $\mathcal{L}_V = \overline{S}_0 \times \tilde{S}_1 \times \cdots \tilde{S}_\Lambda$, it suffices to consider the dualistic structure of the product of *two* manifolds.

Given two Riemannian manifolds $(M, g_1)$, $(N, g_2)$, and points $x_1 \in M$ and $x_2 \in N$. The tangent spaces of $M \times N$ can be expressed as: $T_{(x_1,x_2)}M \times N = T_{x_1}M \oplus T_{x_2}N$. The product Riemannian metric on $M \times N$ is therefore given by $g := g_1 \oplus g_2$ [Lee06]:

$$g\left(X_1|_{x_1} + X_2|_{x_2}, Y_1|_{x_1} + Y_2|_{x_2}\right)|_{(x_1,x_2)} = g_1\left(X_1|_{x_1}, Y_1|_{x_1}\right)|_{x_1} + g_2\left(X_2|_{x_2}, Y_2|_{x_2}\right)|_{x_2},$$

where $X_1|_{x_1}, Y_1|_{x_1} \in T_{x_1}M$ and $X_2|_{x_2}, Y_2|_{x_2} \in T_{x_2}N$. For the rest of the discussion, when the context is clear, we abbreviate the above equation as follows:

$$g\left(X_1 + X_2, Y_1 + Y_2\right) = g_1\left(X_1, Y_1\right) + g_2\left(X_2, Y_2\right), \quad (6.15)$$

where $X_1, Y_1 \in \mathcal{E}(TM)$ and $X_2, Y_2 \in \mathcal{E}(TN)$. Furthermore, suppose $\nabla^1$ and $\nabla^2$ are connections of $M$, $N$ respectively, then the product connection $\nabla$ on $M \times N$ is given by [dC92]:

$$\nabla_{Y_1+Y_2} X_1 + X_2 = \nabla^1_{Y_1} X_1 + \nabla^2_{Y_2} X_2, \quad (6.16)$$

where $X_1, Y_1 \in \mathcal{E}(TM)$ and $X_2, Y_2 \in \mathcal{E}(TN)$. We abbreviate the product connection to a more compact notation for simplicity: $\nabla = \nabla^1 \oplus \nabla^2$. Since the Lie bracket of $M \times N$ is given by:

$$[X_1 + X_2, Y_1 + Y_2]_{M \times N} = [X_1, Y_1]_M + [X_2, Y_2]_N,$$

and the curvature tensor on manifolds given by:

$$R(X, Y)Z = \nabla_X \nabla_Y Z - \nabla_Y \nabla_X Z - \nabla_{[X,Y]} Z,$$

the curvature tensor on the product manifold $M \times N$ is thus given by:

$$R(X_1 + X_2, Y_1 + Y_2, Z_1 + Z_2, W_1 + W_2) = R_1(X_1, Y_1, Z_1, W_1) + R_2(X_2, Y_2, Z_2, W_2), \tag{6.17}$$

where $X_1, Y_1, Z_1, W_1 \in \mathcal{E}(TM)$ and $X_2, Y_2, Z_2, W_2 \in \mathcal{E}(TN)$, and $R_1, R_2$ denote the curvature tensor of $M, N$ respectively. Hence if $M$ and $N$ are flat, so is $M \times N$.

Therefore to show that the Riemannian structure derived from the divergence $D$ in Eq. (6.4) (given by Eq. (6.14)) coincides with the product Riemannian structure discussed above, it suffices to show the following result.

**Theorem 6.2** *Let $(M, g_1, \nabla^1, \nabla^{1^*})$, $(N, g_2, \nabla^2, \nabla^{2^*})$ be two smooth manifolds with their corresponding dualistic structures. Consider the product manifold $M \times N$ with product metric $g = g_1 \oplus g_2$ and product connection $\nabla := \nabla^1 \oplus \nabla^2$. Then the connection $\nabla^{1^*} \oplus \nabla^{2^*}$ is $g$-dual to $\nabla$, i.e. $\left(\nabla^1 \oplus \nabla^2\right)^* = \nabla^{1^*} \oplus \nabla^{2^*}$. Furthermore, if $M$ and $N$ are dually flat, then so is $M \times N$.*

***Proof*** Let $(M, g_1, \nabla^1, \nabla^{1^*})$, $(N, g_2, \nabla^2, \nabla^{2^*})$ be two smooth manifolds with their corresponding dualistic structure. Let $\nabla = \nabla^1 \oplus \nabla^2$ denote the product connection on $M \times N$ given by Eq. (6.16) in compact notation. For simplicity, let $\langle \cdot, \cdot \rangle := \langle \cdot, \cdot \rangle_g$, where $g = g_1 \oplus g_2$ denote the product Riemannian metric on $M \times N$.

Let $X_1, Y_1, Z_1 \in \mathcal{E}(TM)$ and $X_2, Y_2, Z_2 \in \mathcal{E}(TN)$. Given points $x_1 \in M$ and $x_2 \in N$, using the natural identification $T_{(x_1,x_2)} M \times N = T_{x_1} M \oplus T_{x_2} N$, we let:

$$\{X, Y, Z\} := \{X_1 + X_2, Y_1 + Y_2, Z_1 + Z_2\} \in \mathcal{E}(T(M \times N))$$

Then by Eq. (6.15) we have:

$$\begin{aligned} \nabla_X \langle Y, Z \rangle = X \langle Y, Z \rangle &= (X_1 + X_2) \langle Y_1 + Y_2, Z_1 + Z_2 \rangle \qquad (6.18) \\ &= X_1 \left( \langle Y_1, Z_1 \rangle + \langle Y_1, Z_2 \rangle + \langle Y_2, Z_1 \rangle + \langle Y_2, Z_2 \rangle \right) \\ &\quad + X_2 \left( \langle Y_1, Z_1 \rangle + \langle Y_1, Z_2 \rangle + \langle Y_2, Z_1 \rangle + \langle Y_2, Z_2 \rangle \right) \quad , \end{aligned} \tag{6.19}$$

where $X_1, Y_1, Z_1 \in \mathcal{E}(TM)$, $X_2, Y_2, Z_2 \in \mathcal{E}(TN)$.

Let $\pi_{*_1} : T(M \times N) \to TM$ and $\pi_{*_2} : T(M \times N) \to TN$ denote the natural projections, we then have:

$$\begin{aligned} \langle Y_1, Z_2 \rangle &= \langle Y_1, \pi_{*_1} Z_2 \rangle_{g_1} + \langle \pi_{*_2} Y_1, Z_2 \rangle_{g_2} \\ &= \langle Y_1, 0 \rangle_{g_1} + \langle 0, Z_2 \rangle_{g_2} = 0, \end{aligned} \tag{6.20}$$

and by symmetry we obtain $\langle Y_2, Z_1 \rangle = 0$ as well.

Since $\langle Y_1, Z_1\rangle$ is a function on $M$, we have $\nabla^2_{X_2}\langle Y_1, Z_1\rangle = 0$, and by symmetry we obtain $\nabla^1_{X_1}\langle Y_2, Z_2\rangle = 0$ as well. Equation (6.18) therefore becomes:

$$\begin{aligned}
&X_1\left(\langle Y_1, Z_1\rangle + \langle Y_2, Z_2\rangle\right) + X_2\left(\langle Y_1, Z_1\rangle + \langle Y_2, Z_2\rangle\right)\\
&= \nabla^1_{X_1}\langle Y_1, Z_1\rangle + \nabla^1_{X_1}\langle Y_2, Z_2\rangle + \nabla^2_{X_2}\langle Y_1, Z_1\rangle + \nabla^2_{X_2}\langle Y_2, Z_2\rangle\\
&= \nabla^1_{X_1}\langle Y_1, Z_1\rangle + 0 + 0 + \nabla^2_{X_2}\langle Y_2, Z_2\rangle\\
&= \nabla^1_{X_1}\langle Y_1, Z_1\rangle + \nabla^2_{X_2}\langle Y_2, Z_2\rangle. \qquad (6.21)
\end{aligned}$$

By a similar argument, we obtain from Eqs. (6.16) and (6.15):

$$\begin{aligned}
\langle \nabla_X Y, Z\rangle &= \langle \nabla^1_{X_1} Y_1, Z_1 + Z_2\rangle + \langle \nabla^2_{X_2} Y_2, Z_1 + Z_2\rangle\\
&= \langle \nabla^1_{X_1} Y_1, Z_1\rangle + \langle \nabla^1_{X_1} Y_1, Z_2\rangle + \langle \nabla^2_{X_2} Y_2, Z_1\rangle + \langle \nabla^2_{X_2} Y_2, Z_2\rangle\\
&= \langle \nabla^1_{X_1} Y_1, Z_1\rangle + \langle \nabla^2_{X_2} Y_2, Z_2\rangle. \qquad (6.22)
\end{aligned}$$

The first equality is due to Eq. (6.16), and the fact that $\nabla$ is a product connection. The last equality is due to the same argument as in Eq. (6.20): since $\pi_{*_i} Z_j = 0$, and $\pi_{*_i}\nabla^j_{X_j} Y_j = 0$ for $i \neq j$, we have $\langle \nabla^1_{X_1} Y_1, Z_2\rangle = \langle \nabla^2_{X_2} Y_2, Z_1\rangle = 0$.

Finally, subtracting Eq. (6.22) from Eq. (6.21), we have the following equality:

$$\begin{aligned}
X\langle Y, Z\rangle - \langle \nabla_X Y, Z\rangle &= \nabla^1_{X_1}\langle Y_1, Z_1\rangle + \nabla^2_{X_2}\langle Y_2, Z_2\rangle - \left(\langle \nabla^1_{X_1} Y_1, Z_1\rangle + \langle \nabla^2_{X_2} Y_2, Z_2\rangle\right)\\
&= \left(\nabla^1_{X_1}\langle Y_1, Z_1\rangle - \langle \nabla^1_{X_1} Y_1, Z_1\rangle\right) + \left(\nabla^2_{X_2}\langle Y_2, Z_2\rangle - \langle \nabla^2_{X_2} Y_2, Z_2\rangle\right)\\
&= \langle Y_1, \nabla^{1^*}_{X_1} Z_1\rangle + \langle Y_2, \nabla^{2^*}_{X_2} Z_2\rangle,
\end{aligned}$$

where $\nabla^{1^*}, \nabla^{2^*}$ denote the $g_1, g_2$-dual connection to $\nabla^1, \nabla^2$ on $M, N$ respectively. The unique [NS94] $g$-dual connection to $\nabla = \nabla^1 \oplus \nabla^2$ of $M \times N$, denoted by $\nabla^* := \nabla^{1^*} \oplus \nabla^{2^*}$, is thus given by the following:

$$\nabla^*_{Y_1+Y_2} X_1 + X_2 = \nabla^{1^*}_{Y_1} X_1 + \nabla^{2^*}_{Y_2} X_2,$$

where $X_1, Y_1 \in \mathcal{E}(TM)$ and $X_2, Y_2 \in \mathcal{E}(TN)$.

Furthermore, since the curvature of $M \times N$ satisfies the product curvature tensor described in Eq. (6.17), if $(M, g_1, \nabla^1, \nabla^{1^*})$, $(N, g_2, \nabla^2, \nabla^{2^*})$ are both dually flat, then so is their product $(M \times N, g_1 + g_2, \nabla^1 \oplus \nabla^2, \nabla^{1^*} \oplus \nabla^{2^*})$.

□

By Theorems 6.1 and 6.2, the family of parametrized mixture densities $\mathcal{L}_V = \overline{S}_0 \times \tilde{S}_1 \times \cdots \tilde{S}_\Lambda = \overline{S}_0 \times \bigoplus_{\alpha=1}^{\Lambda} \tilde{S}_\alpha$ is therefore a product manifold with product dualistic structure:

$$\left\{g_0 \oplus \bigoplus_{\alpha=1}^{\Lambda} \tilde{g}_\alpha, \nabla^0 \oplus \bigoplus_{\alpha=1}^{\Lambda} \tilde{\nabla}^\alpha, \nabla^{0^*} \oplus \bigoplus_{\alpha=1}^{\Lambda} \tilde{\nabla}^{\alpha^*}\right\}$$

$$= \left\{g^{D_0} \oplus \bigoplus_{\alpha=1}^{\Lambda} g^{\overline{D}_\alpha}, \nabla^{D_0} \oplus \bigoplus_{\alpha=1}^{\Lambda} \nabla^{\overline{D}_\alpha}, \nabla^{D_0^*} \oplus \bigoplus_{\alpha=1}^{\Lambda} \nabla^{\overline{D}_\alpha^*}\right\}.$$

The closure of simplex of mixture coefficients $\overline{S}_0$ can be endowed with a dually flat dualistic structure (Remark 6.2 and Sect. 8.3). By Corollary 5.1, if the mixture component families $S_1, \dots, S_\Lambda$ over $W_1, \dots, W_\Lambda$ in orientation-preserving open cover $\{(\rho_\alpha, W_\alpha)\}_{\alpha=1}^{\Lambda}$ over totally bounded $V \subset M$ are all dually flat, then so are $\tilde{S}_1, \dots, \tilde{S}_\Lambda$. In particular, by Theorem 6.2, this implies $\mathcal{L}_V = \overline{S}_0 \times \tilde{S}_1 \times \cdots \tilde{S}_\Lambda$ is also dually flat.

Finally, consider the case where the families of mixture component densities $\left(\overline{S}_0, g_0, \nabla^0, \nabla^{0^*}\right), \left(\tilde{S}_1, \tilde{g}_1, \tilde{\nabla}^1, \tilde{\nabla}^{1^*}\right), \dots, \left(\tilde{S}_\Lambda, \tilde{g}_\Lambda, \tilde{\nabla}^\Lambda, \tilde{\nabla}^{\Lambda^*}\right)$ are all dually flat manifolds with their corresponding dualistic structures. We show that the mixture divergence defined in Eq. (6.4) is in fact the canonical divergence [AN00] of dually flat manifold $\mathcal{L}_V$.

**Theorem 6.3** *If* $\overline{S}_0, S_1, \dots, S_\Lambda$ *are all dually flat, the divergence function* $D$ *defined in Eq.* (6.4) *is the canonical divergence of* $\mathcal{L}_V = \overline{S}_0 \times \tilde{S}_1 \times \cdots \times \tilde{S}_\Lambda$ *with respect to product dually flat dualistic structure:*

$$\left\{g_0 \oplus \bigoplus_{\alpha=1}^{\Lambda} \tilde{g}_\alpha, \nabla^0 \oplus \bigoplus_{\alpha=1}^{\Lambda} \tilde{\nabla}^\alpha, \nabla^{0^*} \oplus \bigoplus_{\alpha=1}^{\Lambda} \tilde{\nabla}^{\alpha^*}\right\}.$$

***Proof*** Let $(\theta^0, \eta_0)$ denote the local $g_0$-dual coordinates of $\overline{S}_0$, and let $\left(\tilde{\theta}^\alpha\right)$ denote the local $\tilde{\nabla}^\alpha$-affine coordinates of $\tilde{S}_\alpha$ for $\alpha \in [1, \dots, \Lambda]$.

By Theorem 5.2, let $\tilde{\psi}_\alpha$ denote the pulled-back potential function on $\tilde{S}_\alpha$ defined by local coordinates $\left(\tilde{\theta}^\alpha\right)$ and pulled-back metric $\tilde{g}_\alpha$ on $\tilde{S}_\alpha$. The $\tilde{g}_\alpha$-dual local coordinates to $\left(\tilde{\theta}^\alpha\right)$ can be defined via induced potential function $\tilde{\psi}_\alpha$ by: $\left(\tilde{\eta}_\alpha^i := \frac{\partial}{\partial \theta_i^\alpha} \tilde{\psi}_\alpha\right)$.

Since $\overline{S}_0$ and $\tilde{S}_\alpha$ are all dually flat for $\alpha \in [1, \dots, \Lambda]$, we can write the divergences $D_0$ and $\overline{D}_\alpha$ of $\overline{S}_0$ and $\tilde{S}_\alpha$ in the canonical form [AN00] respectively as follows:

$$D_0\left(\{\varphi_\alpha\}_{\alpha=1}^{\Lambda}, \{\varphi'_\alpha\}_{\alpha=1}^{\Lambda}\right) := \psi_0\left(\{\varphi_\alpha\}_{\alpha=1}^{\Lambda}\right) + \psi_0^\dagger\left(\{\varphi'_\alpha\}_{\alpha=1}^{\Lambda}\right) - \left\langle \theta^0\left(\{\varphi_\alpha\}_{\alpha=1}^{\Lambda}\right), \eta_0\left(\{\varphi'_\alpha\}_{\alpha=1}^{\Lambda}\right)\right\rangle,$$

$$\overline{D}_\alpha(\tilde{p}_\alpha, \tilde{q}_\alpha) := \tilde{\psi}_\alpha(\tilde{p}_\alpha) + \tilde{\psi}_\alpha^\dagger(\tilde{q}_\alpha) - \left\langle \tilde{\theta}^\alpha(\tilde{p}_\alpha), \tilde{\eta}_\alpha(\tilde{q}_\alpha)\right\rangle,$$

where $\psi_0$ denotes the canonical divergence on dually flat manifold $\overline{S}_0$ (see for example Bregman divergence defined in Eq. (8.20)).

For $\alpha = 1, \ldots, \Lambda$, the functions $\psi_0^\dagger$ and $\tilde{\psi}_\alpha^\dagger$ denote the Legendre-Fenchel transformation of $\psi_0$ and $\tilde{\psi}_\alpha$ given by the following equations, respectively:

$$\psi_0^\dagger\left(\{\varphi'_\alpha\}_{\alpha=1}^\Lambda\right) := \sup_{\varphi\in\overline{S}_0}\left\{\left\langle\theta^0\left(\{\varphi_\alpha\}_{\alpha=1}^\Lambda\right), \eta_0\left(\{\varphi'_\alpha\}_{\alpha=1}^\Lambda\right)\right\rangle - \psi_0\left(\{\varphi_\alpha\}_{\alpha=1}^\Lambda\right)\right\},$$

$$\tilde{\psi}_\alpha^\dagger(\tilde{q}_\alpha) := \sup_{\tilde{p}_\alpha\in\tilde{S}_\alpha}\left\{\langle\tilde{\theta}^\alpha(\tilde{p}_\alpha), \tilde{\eta}_\alpha(\tilde{q}_\alpha)\rangle - \tilde{\psi}_\alpha(\tilde{p}_\alpha)\right\}.$$

The divergence $D$ on $\mathcal{L}_V$ from Eq. (6.4) is thus given by:

$$\begin{aligned}
D(\tilde{p},\tilde{q}) &= D\left(\sum_{\alpha=1}^\Lambda \varphi_\alpha\tilde{p}_\alpha, \sum_{\alpha=1}^\Lambda \varphi'_\alpha\tilde{q}_\alpha\right) = D_0\left(\{\varphi_\alpha\}_{\alpha=1}^\Lambda, \{\varphi'_\alpha\}_{\alpha=1}^\Lambda\right) + \sum_{\alpha=1}^\Lambda \overline{D}_\alpha(\tilde{p}_\alpha,\tilde{q}_\alpha)\\
&= \psi_0\left(\{\varphi_\alpha\}_{\alpha=1}^\Lambda\right) + \psi_0^\dagger\left(\{\varphi'_\alpha\}_{\alpha=1}^\Lambda\right) - \left\langle\theta^0\left(\{\varphi_\alpha\}_{\alpha=1}^\Lambda\right), \eta_0\left(\{\varphi'_\alpha\}_{\alpha=1}^\Lambda\right)\right\rangle\\
&\quad + \sum_{\alpha=1}^\Lambda\left(\tilde{\psi}_\alpha(\tilde{p}_\alpha) + \tilde{\psi}_\alpha^\dagger(\tilde{q}_\alpha) - \langle\tilde{\theta}^\alpha(\tilde{p}_\alpha), \tilde{\eta}_\alpha(\tilde{q}_\alpha)\rangle\right)\\
&= \left(\psi_0\left(\{\varphi_\alpha\}_{\alpha=1}^\Lambda\right) + \sum_{\alpha=1}^\Lambda\tilde{\psi}_\alpha(\tilde{p}_\alpha)\right) + \left(\psi_0^\dagger\left(\{\varphi'_\alpha\}_{\alpha=1}^\Lambda\right) + \sum_{\alpha=1}^\Lambda\tilde{\psi}_\alpha^\dagger(\tilde{q}_\alpha)\right)\\
&\quad - \left(\left\langle\theta^0\left(\{\varphi_\alpha\}_{\alpha=1}^\Lambda\right), \eta_0\left(\{\varphi'_\alpha\}_{\alpha=1}^\Lambda\right)\right\rangle + \sum_{\alpha=1}^\Lambda\langle\tilde{\theta}^\alpha(\tilde{p}_\alpha), \tilde{\eta}_\alpha(\tilde{q}_\alpha)\rangle\right).
\end{aligned} \tag{6.23}$$

The first and second part of $D$ in Eq. (6.23) are convex due to linearity of derivative, the independence of parameters given by Condition 2, and the fact that the Hessians of potential functions $\psi_0, \tilde{\psi}_1, \ldots, \tilde{\psi}_\Lambda$ are positive semi-definite. The third part of $D$ in Eq. (6.23) is a sum of inner products, which is again an inner product on the product parameter space in $\Xi_0 \times \Xi_1 \times \cdots \times \Xi_\Lambda$.

To show that $D$ is the canonical divergence of $\mathcal{L}_V$, it remains to show that the Legendre–Fenchel transformation of the first component of $D$ in Eq. (6.23) coincide with the second component in Eq. (6.23).

Recall that the parameters of the mixture distribution $\tilde{p}(x,\xi) \in \mathcal{L}_V$ are collected in $(\xi_i)_{i=1}^d := \left(\theta^0, \tilde{\theta}^1, \ldots, \tilde{\theta}^\Lambda\right) \in \Xi_0 \times \Xi_1 \times \cdots \times \Xi_\Lambda$, where $d := \dim(\Xi_0 \times \Xi_1 \times \cdots \times \Xi_\Lambda)$.[5] By linearity of derivative and Condition 2, the $g^D$-coordinate dual to $(\xi_i)_{i=1}^d$ is just the dual coordinates of $\overline{S}_0, \tilde{S}_1, \ldots, \tilde{S}_\Lambda$ expressed in product form: $(\eta_0, \tilde{\eta}_1, \ldots, \tilde{\eta}_\Lambda) = \left(\frac{\partial}{\partial\xi_i}\left(\psi_0\left(\{\varphi_\alpha\}_{\alpha=1}^\Lambda\right) + \sum_{\alpha=1}^\Lambda\tilde{\psi}_\alpha(\tilde{p}_\alpha)\right)\right)$.

For simplicity, we let $\langle\theta^0(\varphi), \eta_0(\varphi')\rangle := \left\langle\theta^0\left(\{\varphi_\alpha\}_{\alpha=1}^\Lambda\right), \eta_0\left(\{\varphi'_\alpha\}_{\alpha=1}^\Lambda\right)\right\rangle$. The dual potential of $\left(\psi_0\left(\{\varphi_\alpha\}_{\alpha=1}^\Lambda\right) + \sum_{\alpha=1}^\Lambda\tilde{\psi}_\alpha(\tilde{p}_\alpha)\right)$, the first part of $D$ in Eq. (6.23), is thus given by the following Legendre-Fenchel transformation:

[5] Recall that coordinates $\theta^0$ of mixture coefficients are not pulled-back.

$$\left(\psi_0\left(\{\varphi_\alpha\}_{\alpha=1}^{\Lambda}\right)+\sum_\alpha\tilde{\psi}_\alpha(\tilde{p}_\alpha)\right)^\dagger =$$

$$\sup_{\varphi\in\overline{S}_0,\tilde{p}_\alpha\in\tilde{S}_\alpha}\left\{\left(\left\langle\theta^0(\varphi),\eta_0(\varphi')\right\rangle+\sum_{\alpha=1}^{\Lambda}\left\langle\tilde{\theta}^\alpha(\tilde{p}_\alpha),\tilde{\eta}_\alpha(\tilde{q}_\alpha)\right\rangle\right)-\left(\psi_0\left(\{\varphi_\alpha\}_{\alpha=1}^{\Lambda}\right)+\sum_\alpha\tilde{\psi}_\alpha(\tilde{p}_\alpha)\right)\right\}$$

$$=\sup_{\varphi\in\overline{S}_0,\tilde{p}_\alpha\in\tilde{S}_\alpha}\left\{\left(\left\langle\theta^0(\varphi),\eta_0(\varphi')\right\rangle-\psi_0\left(\{\varphi_\alpha\}_{\alpha=1}^{\Lambda}\right)\right)+\sum_{\alpha=1}^{\Lambda}\left(\left\langle\tilde{\theta}^\alpha(\tilde{p}_\alpha),\tilde{\eta}_\alpha(\tilde{q}_\alpha)\right\rangle-\tilde{\psi}_\alpha(\tilde{p}_\alpha)\right)\right\}$$

$$=\sup_{\varphi\in\overline{S}_0}\left\{\left\langle\theta^0(\varphi),\eta_0(\varphi')\right\rangle-\psi_0\left(\{\varphi_\alpha\}_{\alpha=1}^{\Lambda}\right)\right\}+\sum_{\alpha=1}^{\Lambda}\sup_{\tilde{p}_\alpha\in\tilde{S}_\alpha}\left\{\left\langle\tilde{\theta}^\alpha(\tilde{p}_\alpha),\tilde{\eta}_\alpha(\tilde{q}_\alpha)\right\rangle-\tilde{\psi}_\alpha(\tilde{p}_\alpha)\right\}$$

$$=\psi_0^\dagger\left(\{\varphi'_\alpha\}_{\alpha=1}^{\Lambda}\right)+\sum_{\alpha=1}^{\Lambda}\tilde{\psi}_\alpha^\dagger(\tilde{q}_\alpha).$$

The third equality follows from the functional independence of $\{\varphi_\alpha\}_{\alpha=1}^{\Lambda}\in\overline{S}_0$ and $\tilde{p}_\alpha$'s in $\tilde{S}_\alpha$. Hence the Legendre-Fenchel transform of the first component of $D$ is exactly the second component of $D$, and $D$ is the canonical divergence of the family of mixture densities, given by:

$$\left(\mathcal{L}_V, g_0\oplus\bigoplus_{\alpha=1}^{\Lambda}\tilde{g}_\alpha, \nabla^0\oplus\bigoplus_{\alpha=1}^{\Lambda}\tilde{\nabla}^\alpha, \nabla^{0^*}\oplus\bigoplus_{\alpha=1}^{\Lambda}\tilde{\nabla}^{\alpha^*}\right).$$

□

## 6.5 Towards a Population-Based Optimization Method on Riemannian Manifolds

Inspired by the geometrization of statistical models in Information Geometry and the recent developments of manifold optimization methods, the first part of the book constructs a geometrical framework for stochastic optimization on Riemannian manifolds that combines the statistical geometry of the decision space and the Riemannian geometry of the search space.

In the previous chapter, we described a notion of family of parametrized local densities on Riemannian manifold $M$ locally inherited from Euclidean space. By viewing the locally inherited probability densities on $M$ as probability $n$-forms, we showed that a dualistic statistical geometry on family of local densities can be inherited entirely from a family of probability densities on Euclidean spaces. This relates the statistical parameters of the statistical manifold decision space and the *local* point estimations on $M$; however, the local restriction still persisted.

In this chapter, we extended the notion of locally inherited probability densities beyond the normal neighbourhood and constructed a family of parametrized mixture densities $\mathcal{L}_V$ on totally bounded subsets $V$ of $M$. We derived the geometrical structure of the family of mixture densities $\mathcal{L}_V$ and showed that $\mathcal{L}_V$ is a product statistical manifold of the simplex of mixture coefficients and the families of mixture component densities.

The significance of the product structure of $\mathcal{L}_V$ is twofold: it provides a computable geometrical structure for finitely parametrized statistical model on $M$ that extends beyond the confines of a single normal neighbourhood, and allows us to handle statistical parameter estimations and computations of mixture coefficients and mixture components independently.

The product Riemannian structure of mixture densities $\mathcal{L}_V$ over Riemannian manifolds thus provides us with a geometrical framework to tackle the optimization problem described in Sect. 1.2. In the second part of the book we apply this framework to construct a population-based derivative-free meta-algorithm on Riemannian manifolds using the statistical geometry of the decision space and the Riemannian geometry of the search space, overcoming the locality of manifold optimization algorithms in the literature.

## References


[AN00] S. Amari and H Nagaoka. *Methods of Information Geometry*, volume 191 of *Translations of Mathematical monographs*. Oxford University Press, 2000.

[BB08] Abhishek Bhattacharya and Rabi Bhattacharya. Statistics on riemannian manifolds: asymptotic distribution and curvature. *Proceedings of the American Mathematical Society*, 136(8):2959–2967, 2008.

[CU14] Ovidiu Calin and Constantin Udrişte. *Geometric modeling in probability and statistics*. Springer, 2014.

[dC92] M.P. do Carmo. *Riemannian Geometry*. Mathematics (Boston, Mass.). Birkhäuser, 1992.

[E+92] Shinto Eguchi et al. Geometry of minimum contrast. *Hiroshima Mathematical Journal*, 22(3):631–647, 1992.

[Lee06] John M Lee. *Riemannian manifolds: an introduction to curvature*, volume 176. Springer Science & Business Media, 2006.

[NS94] Katsumi Nomizu and Takeshi Sasaki. *Affine differential geometry: geometry of affine immersions*. Cambridge University Press, 1994.

[Oll93] Josep M Oller. On an intrinsic analysis of statistical estimation. In *Multivariate Analysis: Future Directions 2*, pages 421–437. Elsevier, 1993.

[Pen04] Xavier Pennec. *Probabilities and statistics on Riemannian manifolds: A geometric approach*. PhD thesis, INRIA, 2004.

[Sak96] Takashi Sakai. *Riemannian Geometry*, volume 149. American Mathematical Soc., 1996.

# Part II
# Model-Based Stochastic Derivative-Free Optimization on Riemannian Manifolds

# Chapter 7
# Geometry in Optimization

**Abstract** Equipped with the product statistical manifold structure of mixture densities, the second part of the book begins by surveying the geometric aspects of two contemporary branches of optimization theories. We first review adaptations of optimization algorithms from Euclidean spaces to Riemannian manifolds, otherwise known as manifold optimization or Riemannian optimization in the literature. We then discuss the information geometric interpretation of population-based stochastic optimization algorithms on Euclidean spaces. We show that Riemannian adaptations of Euclidean optimization algorithms to Riemannian manifolds all effectively employ the same principle as the "statistical" approach described in Chap. 4. As a result the locality and assumptions of the "statistical" approach still persisted in manifold optimization algorithms, which in turn limits the generality of manifold optimization algorithms in the literature. In order to overcome the locality and implicit assumptions of the Riemannian adaptation process, we require parametrized probability densities defined beyond the confines of a single normal neighbourhoods on Riemannian manifolds, and the mixture densities described in Chap. 6 provides exactly what is needed. This leads us to the discussion of the next chapter, where we overcome the local restrictions by proposing a population-based meta-algorithm using the geometrical framework described in the first part of the book.

In this chapter we survey the geometric aspects of two contemporary branches of optimization theories described in Sect. 1.1. We first review adaptations of optimization algorithms from Euclidean spaces to Riemannian manifolds, otherwise known as manifold optimization or Riemannian optimization in the literature [AH19]. We then discuss the information geometric interpretation of population-based stochastic optimization algorithms on Euclidean spaces [MMP11, ANOK12, WSPS08a, OAAH17].

Manifold optimization originated from the observation that data measurements on Euclidean spaces typically admit an underlying geometrical structure (known as manifold hypothesis [FMN16]). Since the dimension of the underlying data manifold is lower than that of the ambient Euclidean space, optimization problems on the data manifold can be solved with lower complexity using approaches *intrinsic* to the underlying data manifold.

 
R. S. Fong and P. Tino, *Population-Based Optimization on Riemannian Manifolds*,
Studies in Computational Intelligence 1046,
https://doi.org/10.1007/978-3-031-04293-5_7

This motivates the notion of manifold optimization—the optimization on the data manifold search space using only its intrinsic geometrical properties. That is, the manifold search space is viewed as a "stand-alone" non-linear search space, free from any ambient Euclidean space.

Since optimization techniques on Euclidean spaces are well studied and well established, it is natural to adapt and translate existing optimization methods on Euclidean spaces to the context of Riemannian manifolds. Indeed, this is the approach of Riemannian adaptations of optimization algorithms in the literature, which effectively employ the *same* principle as the "statistical" approach (of constructing probability distribution on manifolds), described in Chap. 4: first identify the *necessary* ingredients of existing optimization algorithms and re-establish them in the manifold context. As a result, the locality of the "statistical approach" persists. Furthermore, since the adaptation process depends *solely* on the Riemannian exponential map, additional assumptions have be to made to accommodate the adaptation process. This, in turn limits the generality of manifold optimization algorithms.

On the other hand, recent advancements in information geometrical provides us with geometrical insights into the decision space of population-based stochastic optimization methods on Euclidean spaces. By viewing statistical models over Euclidean spaces as Riemannian manifolds, the evolution of sampling distributions in stochastic optimization algorithms can be viewed as search trajectory on the overarching statistical manifold.

However, a direct adaptation of this geometrical view point to the context of manifold optimization only further accentuates the locality of the Riemannian adaptation approach. In order to overcome the local restrictions of the Riemannian adaptation approach, we require parametrized probability densities defined beyond the confines of a single normal neighbourhoods on Riemannian manifolds, and the mixture densities described in Chap. 6 provides exactly what is needed.

This leads us to the discussion of the next chapter, where we first accentuate the local restrictions of the Riemannian adaptation approach by first generalizing population-based Riemannian stochastic optimization algorithms, and then overcome the local restrictions by proposing a population-based meta-algorithm using the geometrical framework described in Part I.

The remainder of the chapter will be organized as follows.

In Sect. 7.1, we discuss the general principle of translating optimization algorithms from Euclidean spaces to Riemannian manifolds in the literature. The principle is the same as the "statistical" approach of Chap. 4, whilst the approach differs only by additional ingredients. In this section we describe the general principle, the implicit assumptions, and the additional arithmetic ingredients formally.

In Sect. 7.2, we review three Riemannian adaptations of optimization algorithms in the literature derived from the principle described in Sect. 7.1. In particular, we consider Riemannian adaptations of gradient-based method (Trust Region) [ABG07], meta-heuristic Particle Swarm Optimization (PSO) [BIA10] and stochastic derivative-free optimization (Covariance Matrix Adaptation Evolution Strategy (CMA-ES)) [CFFS10].

In Sect. 7.3, we discuss the Information Geometric interpretation of population-based stochastic derivative-free optimization (SDFO), such as CMA-ES, over Euclidean spaces. We discuss the local restrictions of the Riemannian adaptation approach described in Sect. 7.1, and how Riemannian CMA-ES does not share the same property as its Euclidean counterpart.

## 7.1 Principle of Riemannian Adaptation

The procedure of Riemannian adaptation of optimization algorithms in the literature can be summarized as follows: for each iteration, the optimization method determines a search direction on the tangent space centered at the current search iterate. The next search iterate is then obtained by tracing along the *local* geodesic from the current iterate along the search direction via the Riemannian exponential map.[1] The *local* search information is subsequently parallel transported to (the tangent space of) the new search iterate. The decision space of Riemannian adapted algorithms is thus given by the *local, disjoint* tangent (Euclidean) spaces, whereas the search space is given by the Riemannian manifold.

The computation and estimation of the search direction on each search iterate are therefore 'out-sourced' to (the pre-image of normal neighbourhood in) the tangent space, which is a vector space. That is, the local computations on the search space manifold of the Riemannian adapted optimization algorithms are performed on the tangent spaces in the same fashion as the pre-adapted Euclidean counterparts. However, this also means the aforementioned computations and estimations are strictly local, as they are confined within a *single* normal neighbourhood around the search iterates (a subset of the local disjoint tangent spaces).

The above discussion is summarized in Fig. 7.1 below.

The locality of the normal neighbourhood in turn gives raise to the implicit assumptions of manifold optimization algorithms in the literature. To guarantee that a new search iterate on the search space manifold can be obtained through any generated search direction on the tangent space, we would require one of the following:

1. Local search directions are generated within the pre-image of a normal neighbourhood, or
2. The Riemannian exponential map is defined on the entire tangent spaces $T_x M$ for all $x \in M$.

Whilst the former enforces locality on the optimization process, the latter restricts the generality of the optimization algorithm by making additional assumptions on the search space manifold. Similar to the statistical approach (Sect. 4.2), manifold optimization methods in the literature chose the latter, and assume the search space

[1] Gradient-based algorithms can have a slightly relaxed variations of this map called retraction [AMS09], as we will discuss in Sect. 7.2.1. The principle concept is the same.

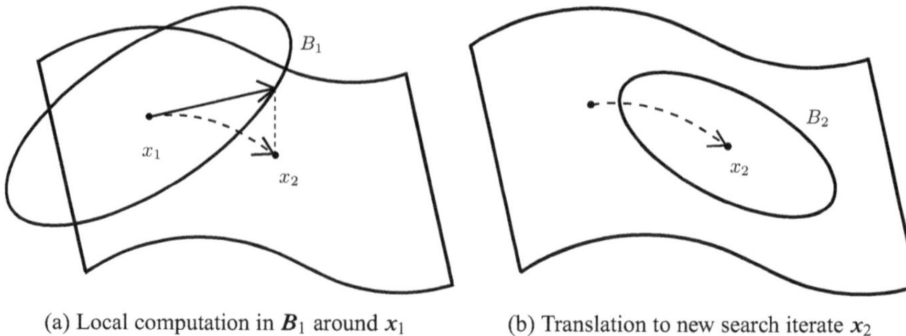


(a) Local computation in $\boldsymbol{B}_1$ around $\boldsymbol{x}_1$ (b) Translation to new search iterate $\boldsymbol{x}_2$

**Fig. 7.1** Illustration of Riemannian adaptation of optimization algorithms. For each iteration, Riemannian optimization initiate at $x_1 \in M$, the local search direction is obtained by performing local computation on $B_1 \subset T_{x_1}M$. A new search iterate $x_2$ is obtained by tracing along the geodesic defined by the search direction illustrated in Fig. 7.1a. This process is repeated for $x_2$ on $B_2 \subset T_{x_2}M$ in Fig. 7.1b

Riemannian manifold to be **complete**. Hopf-Rinow theorem then asserts the Riemannian exponential map is defined for all elements of any tangent space $T_xM$. The drawbacks of this assumption will be further discussed in Chap. 9.

This principle has since been adopted by authors to translate optimization methods from Euclidean space to **complete** Riemannian manifolds.[2] Notable examples include gradient-based methods on matrix manifolds [Gab82, AMS09], metaheuristics (derivative-free methods) such as Particle Swarm Optimization (PSO) on complete Riemannian manifolds [BIA10] and stochastic derivative-free optimization algorithms such as Covariance Matrix Adaptation Evolutionary Strategies (CMA-ES) [HO96, KMH+04, Han06] on spherical manifolds [CFFS10].

Similar to the "statistical" approach described in Chap. 4, it remains to find the necessary arithmetic operations of optimization algorithms on Euclidean spaces, and determine the analogous notion on manifolds. Table 7.1 below summarizes the necessary ingredients used in the principle, extending Table 4.1 described in Chap. 4:

For the remainder of the section, we discuss formally the additional ingredients listed in the last three rows of Table 7.1.

### 7.1.1 Riemannian Gradient and Hessian

Given a Riemannian manifold $(M, g)$ and affine connection $\nabla$ on $M$. Let $f : M \to \mathbb{R}$ be a smooth real-valued function on $M$. For any point $x \in M$ the **Riemannian gradient** of $f$ at $x$, denoted by grad $f(x) \in T_xM$, is the tangent vector at $x$ *uniquely* associated to the differential of $f$. In particular, grad $f(x)$ is the unique tangent vector satisfying:

[2] The authors assume completeness of the search space manifold for the same reason as the "statistical" approach in Chap. 4: to extend the normal neighbourhood as far as possible.

**Table 7.1** Conversion of arithmetic operation for Riemannian adaption of optimization algorithms

| | Euclidean spaces | Riemannian manifold |
|---|---|---|
| + | $x + v$ | $\exp_x(v)$ |
| − | $y - x =: \mathbf{xy}$ | $\mathbf{xy} = \exp_x^{-1}(y)$ |
| Distance | $\mathrm{dist}(x, y) := \|\|x - y\|\|$ | $\mathrm{dist}(x, y) = \sqrt{\langle \mathbf{xy} \rangle_g}$ |
| Gradient | $\nabla f$ | Riemannian gradient, "natural gradient" |
| Hessian | Hess $f$ | Riemannian Hessian |
| Translation of search information | n/a | Parallel transport (within normal neighbourhood) |

$$\langle \mathrm{grad}\, f(x), v \rangle_g = df(v)|_x, \quad v \in T_x M.$$

The Riemannian gradient grad $f \in \mathcal{E}(TM)$ is therefore a vector field on $M$. In local coordinates $\left(x^1, \ldots, x^n\right)$ of $M$, Riemannian gradient grad $f$ can be expressed as:

$$\mathrm{grad}\, f = \sum_{i,j=1}^{n} g^{ij} \frac{\partial f}{\partial x^i} \frac{\partial}{\partial x^j},$$

where $\left[g^{ij}\right]$ denote the inverse of Riemannian metric matrix associated to $g$ and $\left(\frac{\partial}{\partial x^i}\right)_{i=1}^{n}$ denote the local coordinate frame corresponding to the local coordinate system $\left(x^1, \ldots, x^n\right)$ described in Sect. 2.2. Furthermore, for each $x \in M$, we may associate grad $f$ to a column vector for each $x \in M$:

$$\tilde{\nabla} f(x) := \sum_{i=1}^{n} g^{ij} \frac{\partial f}{\partial x^i} = G^{-1}(x) \nabla f(x), \tag{7.1}$$

where $G$ denote the Riemannian metric matrix associated to $g$ and $\nabla f(x)$ denote the gradient in Euclidean spaces from elementary calculus. The vector $\tilde{\nabla} f(x)$ is otherwise known as **natural gradient** [Ama98] in the machine learning community.

On the other hand, the **Riemannian Hessian** of a smooth function $f \in C^\infty(M, \mathbb{R})$ is the $(0, 2)$ tensor field given by:

$$\begin{aligned} \mathrm{Hess}\, f : TM \times TM &\to \mathbb{R} \\ (X, Y) &\mapsto \nabla\nabla f = \langle \nabla_X \mathrm{grad}\, f, Y \rangle_g. \end{aligned}$$

## 7.2 Examples of Riemannian Adaptation of Optimization Algorithms in the Literature

In this section we illustrate the Riemannian adaptation approach with algorithms from the literature: Riemannian adaptations of trust-region method [ABG07], Particle Swarm Optimization (PSO) [BIA10] and Covariance Matrix Adaptation Evolution Strategy (CMA-ES) [CFFS10].

For the remainder of the section, let $M$ denote the Riemannian manifold search space, and let $f : M \to \mathbb{R}$ denote the objective function of a **minimization** problem over $M$. For each $x \in M$, let $\tilde{f}_x := f \circ \exp_x : T_x M \to \mathbb{R}$ denote the locally converted objective function on *each* tangent space $T_x M$ of $M$. For each $x \in M$, let $U_x := \exp_x^{-1}(N_x) \subset T_x M$ denote the neighbourhood centered at $\mathbf{0} \in T_x M$ given by the pre-image of a normal neighbourhood $N_x$ of $x$ under the Riemannian exponential map.

### 7.2.1 *Riemannian Gradient-Based Optimization*

In this section we review Riemannian adaptation [ABG07] of Trust Region method [NW06]. This is a prototypical example of how the Riemannian adaptation approach is applied to translate gradient-based optimization from Euclidean spaces to the context of Riemannian manifolds. Other gradient-based algorithms in the literature are translated under the same principle described in Sect. 7.1 [AMS09]. The following side-by-side comparison of **Euclidean Trust-Region** [NW06] and **Riemannian Trust Region** [ABG07] illustrates the Riemannian adaption process of gradient-based methods.

The hyperparameters and the corresponding numerical values of both versions of Trust-Region, summarized in Table 7.2 below, are taken from the corresponding references [NW06, ABG07]. It is worth nothing that, while the numerical values may differ across different implementations, the purpose of this discussion is to illustrate the Riemannian adaptation process highlighted in the algorithms below.

**Table 7.2** Parameters of trust-region

| Parameter | Value | Meaning |
|---|---|---|
| $\overline{\Delta}$ | $\pi$ | Upperbound on step length |
| $\Delta_0$ | $\Delta_0 \in (0, \Delta)$ | Initial step length |
| $\rho'$ | $\rho' \in \left[0, \frac{1}{4}\right)$ | Parameter that determines whether the new solution is accepted |

**Algorithm 7.1:** Euclidean Trust-Region

**Data**: Initial point $x_0 \in \mathbb{R}^n$, $\Delta_0 \in (0, \Delta)$.

1 **for** $k = 0, 1, 2, \ldots$ **do**
2     Obtain $\eta_k$ by solving **Euclidean** Trust-Region sub-problem. ;
3     Compute reduction ratio $p_k$ ;
4     **if** $p_k < \frac{1}{4}$ **then**
5         $\Delta_{k+1} = \frac{1}{4}\Delta_k$ ;
6         **else if** $p_k > \frac{3}{4}$ *and* $||\eta_k|| = \Delta_k$ **then**
7             $\Delta_{k+1} = \min(2 \cdot \Delta_k, \Delta)$;
8         **else**
9             $\Delta_{k+1} = \Delta_k$ ;
10     **if** $p_k > \rho'$ **then**
11         $x_{k+1} = x_k + \eta_k$ ;
12         **else**
13             $x_{k+1} = x_k$ ;
14

**Algorithm 7.2:** Riemannian Trust-Region

**Data**: Initial point $x_0 \in M$, $\Delta_0 \in (0, \Delta)$.

1 **for** $k = 0, 1, 2, \ldots$ **do**
2     Obtain $\eta_k$ by solving **Riemannian** Trust-Region sub-problem. ;
3     Compute reduction ratio $p_k$ ;
4     **if** $p_k < \frac{1}{4}$ **then**
5         $\Delta_{k+1} = \frac{1}{4}\Delta_k$ ;
6         **else if** $p_k > \frac{3}{4}$ *and* $||\eta_k|| = \Delta_k$ **then**
7             $\Delta_{k+1} = \min(2 \cdot \Delta_k, \Delta)$;
8         **else**
9             $\Delta_{k+1} = \Delta_k$ ;
10     **if** $p_k > \rho'$ **then**
11         $x_{k+1} = \exp_{x_k}(\eta_k)$ ;
12         **else**
13             $x_{k+1} = x_k$ ;
14

The two algorithms differ only in the highlighted lines: the Trust-Region subproblem in line 2, computation of reduction ratio in line 3, and the translation to the new search iterate in line 11. Since we have already discussed the translation of search iterate in the beginning of the chapter (summarized by the addition operation $+$ in Table 7.1), it remains to discuss both the adaptation of trust-region subproblem and reduction ratio from Euclidean spaces to Riemannian manifolds.

On Euclidean spaces, the **Euclidean trust-region subproblem** (line 2) at the current search iterate $x_k$ on $\mathbb{R}^n$ is given by [NW06]:

$$\eta_k := \arg\min_{\eta \in \mathbb{R}^n} m_k(\eta) = \arg\min_{\eta \in \mathbb{R}^n} f(x_k) + \langle \nabla f(x_k), \eta \rangle + \frac{1}{2}\eta^\top H_f(x_k)\eta,$$
$$\text{s.t. } ||\eta_k|| \leq \Delta_k$$

where $\nabla f(x_k)$ is the gradient on Euclidean space, $H_f(x_k)$ is the Hessian matrix of $f$ at $x_k$, and $\langle \cdot, \cdot \rangle$ denote the standard inner product on $\mathbb{R}^n$.

On Riemannian manifold $M$, the **Riemannain trust-region subproblem** is therefore translated *locally* from the Euclidean version, around the search iterate $x_k \in M$, with the tools described in Sect. 7.1:

$$\eta_k := \arg\min_{\eta \in T_{x_k} M} m_{x_k}(\eta)$$
$$= \arg\min_{\eta \in T_{x_k} M} f(x_k) + \langle \text{grad}\, f(x_k), \eta \rangle_g + \frac{1}{2}\text{Hess}\, f(x_k)\,(\eta, \eta)\ , \ \text{s.t. } \sqrt{\langle \eta, \eta \rangle_g} \leq \Delta_k,$$

where grad $f(x_k)$ and Hess $f(x_k)$ $(\eta, \eta)$ denote the Riemannian gradient and Hessian function described in Sect. 7.1.1, and $\langle \cdot, \cdot \rangle_g$ is the Riemannian metric. It is important to note that the above formulation is strictly local within the tangent space $T_{x_k}M$ and can be equivalently formulated as minimization problem on the tangent space with traditional gradient and Hessian of $\tilde{f}$:

$$\begin{aligned}\eta_k &:= \arg\min_{\eta\in T_{x_k}M} m_{x_k}(\eta)\\ &= \arg\min_{\eta\in T_{x_k}M} \tilde{f}(x_k) + \langle\nabla\tilde{f}(x_k),\eta\rangle_g + \frac{1}{2}\eta^\top H_{\tilde{f}}(x_k)\eta\,,\ \text{s.t.}\ \sqrt{\langle\eta,\eta\rangle_g} \le \Delta_k.\end{aligned}$$

Finally, the **translation of reduction ratio** $p_k$ is described as follows:

$$p_k = \underbrace{\frac{f(x_k) - f(x_k+\eta_k)}{m_k(0) - m_k(\eta_k)}}_{\text{Euclidean case}} \longrightarrow \underbrace{\frac{f(x_k) - f\left(\exp_{x_k}(\eta_k)\right)}{m_{x_k}(0) - m_{x_k}(\eta_k)}}_{\text{Riemannian case}}.$$

**Remark 7.1** *(Retraction)* The Riemannian exponential map is generally difficult to compute, authors of [AMS09] proposed a slightly relaxed map from the tangent spaces to $M$ called the retraction.

**Definition 7.1** A **retraction** on a manifold is a (smooth) function $R_x : T_xM \to M$ satisfying the following:

1. $R_x(\mathbf{0}) = x$, where $\mathbf{0} \in T_xM$ and
2. $DR_x(\mathbf{0}) = \mathrm{id}_{T_xM}$ .

The notion of retraction provides the sufficient conditions to locally preserve the gradient information within a normal neighbourhood centered at $x \in M$. This generalization is similar to the generalization of Riemannian exponential map to orientation-preserving diffeomorphism described in Sect. 5.2.

### *7.2.2 Riemannian Particle Swarm Optimization*

In this section, we discuss the Riemannian adaptation of **Particle Swarm Optimization** (PSO) [EK95], a population-based meta-heuristic optimization algorithm, from Euclidean spaces to Riemannian manifolds discussed in [BIA10].

The Riemannian adaptation of PSO is similar to the Riemannian adaptation of trust-region method described in the previous section: The Euclidean version and the Riemannian version differ only in the local computations, while the structure of the algorithm remains the same. In particular, the local estimations and computations involved in the evolution of search iterates is performed on the local pre-image of normal neighbourhoods just as its Euclidean counterparts.

In classical PSO on Euclidean space $\mathbb{R}^n$ discussed in [EK95], search agents (particles) are initially randomly generated on $\mathbb{R}^n$. On the subsequent $k$th iteration each particle $x_i^k$ 'evolves' under the following equation (Eqs. (5) and (6) of [BIA10]):

**Table 7.3** Meaning of symbols in PSO step (Eq. (7.2))

| Symbols | Meaning |
|---|---|
| $x_i^k$ | Position of the $i$th particle at the $i$th iteration |
| $v_i^k$ | Search direction (overall velocity) of the $i$th particle at the $k$th iteration |
| $w^k$ | Inertial coefficient, assigned by a predefined real valued function on the iteration counter $k$ |
| $c$ | Weight of the nostalgia component, a predefined real number |
| $s$ | Weight of the social component, a predefined real number |
| $\alpha_i^k, \beta_i^k$ | Random numbers generate with [0, 1] for each of the $i$th particle at the $k$th iteration of the nostalgia and social components respectively. The purpose of the random components is to avoid premature convergence [EK95] |
| $y_i^k$ | "Personal best" of the $i$th particle up to the $k$th iteration |
| $\hat{y}^k$ | Overall best of all the particles up to the $k$th iteration |

$$v_i^{k+1} = \underbrace{w^k \cdot v_i^k}_{\text{inertia}} + \underbrace{c \cdot \alpha_i^k \left(y_i^k - x_i^k\right)}_{\text{nostalgia}} + \underbrace{s \cdot \beta_i^k \left(\hat{y}^k - x_i^k\right)}_{\text{social}} \tag{7.2}$$

$$x_i^{k+1} = x_i^k + v_i^{k+1}. \tag{7.3}$$

The meaning of the symbols are summarized in Table 7.3.

The adaptation of the second PSO equation (Eq. (7.3)) to Riemannian manifolds is similar to that of the translation of search iterates described in line 11 of Riemannian trust region: using the Riemannian exponential map, we rewrite (see Table 7.1):

$$x_i^{k+1} = x_i^k + v_i^{k+1} \to x_i^{k+1} = \exp_{x_i^k}\left(v_i^{k+1}\right),$$

where the completeness assumption of $M$ is used when $||v_i^{k+1}|| > \operatorname{inj}\left(x_i^k\right)$.

On the other hand, the adaptation of the nostalgia and social component in the first PSO equation (Eq. (7.2)) involves the difference of points on the manifold. In [BIA10], the nostalgia component (similarly the social component) of the first PSO equation is adapted using the Riemannian logarithm map (as described by the subtraction operation—in Table 7.1):

$$c \cdot \alpha_i^k \left(y_i^k - x_i^k\right) \longrightarrow c \cdot \alpha_i^k \cdot \log_{x_i^k}\left(y_i^k\right).$$

However, it is important to note that this construction only exists when the points $y := y_i^k$ and $x := x_i^k$ are "close enough". Indeed, the Riemannian logarithm map, being the inverse of Riemannian exponential map, exists *only* when the Riemannian exponential map is invertible, i.e. when $y$ is within the normal neighborhood of $x$ (and vice versa). Therefore, the full evolutionary step of Riemannian adaptation of PSO *only* works when *all* the points (particles) are within a *single* normal neighbourhood. The assumptions of Riemannian PSO is further discussed in Sect. 9.1.

### 7.2.3 Riemannian CMA-ES

In this section, we review the Riemannian adaptation of **Covariance Matrix Adaptation Evolutionary Strategies** (CMA-ES) [HO96, KMH+04, Han06], illustrated on spherical manifolds in [CFFS10], where further studies of properties of step length were made in [Arn14].

We begin by describing general Stochastic Derivative-Free Optimization (SDFO) algorithms on Euclidean spaces. Let $f : \mathbb{R}^n \to \mathbb{R}$ denote the objective function of an optimization problem over $\mathbb{R}^n$. Consider a parametrized family of probability densities over $\mathbb{R}^n$: $\{p(\cdot|\theta)|\theta \in \Xi \subset \mathbb{R}^m\}$. Examples of SDFO algorithms include Estimation of Distribution Algorithms (EDA) [LL01], Evolutionary Algorithms such as CMA-ES [HO96], and natural gradient optimization methods [OAAH17]. We summarize SDFO algorithms on Euclidean space in Algorithm 7.3 below:

**Algorithm 7.3:** General SDFO on Euclidean spaces

**Data**: Initial point $x_0 \in \mathbb{R}^n$, initial parameter $\theta^0 \in \Xi$

**while** *stopping criterion not satisfied* **do**
- Generate a sample $X^k \sim p(\cdot|\theta^k)$ ;
- Evaluate fitness $f(x)$ of each $x \in X^k$ ;
- Update $\theta^{k+1}$ based on the fittest subset of the sample: $\hat{X}^k \subset X^k$ ;
- k = k+1;

For the remainder of the section we focus our attention on CMA-ES [HO96], a state-of-the-art SDFO algorithm that has been adapted to the context of Riemannian manifolds [CFFS10] using the same principle described in Sect. 7.1.

We begin by summarizing the essential structure of the "vanilla" version of CMA-ES in Algorithm 7.4 below. It is worth noting that while a variety of different implementations of CMA-ES exists and its extensions studied extensively in the literature, the purpose of this section is to illustrate the Riemannian adaptation process highlighted in the algorithms.

The parameters of CMA-ES, described in [Han16, CFFS10], is summarized in Table 7.4.

To translate CMA-ES from Euclidean spaces to Riemannian manifolds, authors of Riemannian adaptation of CMA-ES (RCMA-ES) [CFFS10] implicitly adopted the "statistical" approach of locally inherited probability densities (within normal neighbourhoods) described in Chap. 4, Sect. 5.2.2, and [Pen06]. The adaptation process is therefore similar to the translation of Euclidean trust-region to Riemannian trust-region described in Sect. 7.2.1 above: for each iteration a local search distribu-

**Table 7.4** Parameters of CMA-ES described in [CFFS10]

| Parameter | Default value [Han16] | Meaning |
|---|---|---|
| $m_1$ | Depends on the search space | Number of sample parents per iteration |
| $m_2$ | $\frac{m_1}{4}$ | Number of offsprings. |
| $w_i$ | $\log\left(\frac{m_2+1}{i}\right) \cdot \left(\sum_{j=1}^{m_2} \log\left(\frac{m_2+1}{j}\right)\right)^{-1}$ | Recombination coefficient |
| $m_{\text{eff}}$ | $\left(\sum_{i=1}^{m_2} w_i^2\right)^{-1}$ | Effective samples |
| $c_c$ | $\frac{4}{N+4}$ | Learning rate of anisotropic evolution path |
| $c_\sigma$ | $\frac{m_{\text{eff}}+2}{N+m_{\text{eff}}+3}$ | Learning rate of isotropic evolution path |
| $\mu_{\text{cov}}$ | $m_{\text{eff}}$ | Factor of rank-$m_2$-update of Covariance matrix |
| $c_{\text{cov}}$ | $\frac{2}{\mu_{\text{cov}}\left(N+\sqrt{2}\right)^2} + \left(1 - \frac{1}{\mu_{\text{cov}}}\right) \min\left(1, \frac{2\mu_{\text{cov}}-1}{(N+2)^2+\mu_{\text{cov}}}\right)$ | Learning rate of covariance matrix update |
| $d_\sigma$ | $1 + 2\max\left(0, \sqrt{\frac{m_{\text{eff}}-1}{N+1}}\right) + c_\sigma$ | Damping parameter |

tion is generated on the normal neighbourhood centered at the search iterate. The computations and estimation of search information, such as estimation covariance matrix $C$, is performed locally within the pre-image of the normal neighbourhood. The results are subsequently parallel transported to the new search iterate via the Riemannian exponential map. The structure of the Riemannian adapted algorithm remains the the same as the pre-adapted algorithm on Euclidean spaces. Therefore, it remains to describe the changes in sampling process and translation to new search center highlighted in Algorithm 7.4.

For each $k > 0$, at the $k$th iteration of RCMA-ES, a new set of sample vectors is generated from the tangent space $T_{\mu_k}M$ within the pre-image of normal neighbourhood $N_{\mu_k}$ around the search iterate $\mu_k \in M$:

$$\left\{v_{i,\mu_k}\right\}_{i=1}^{m_1} \subset T_{\mu_k}M \sim N\left(\mathbf{0}, (\sigma)^2 \cdot C\right), \quad \text{s.t. } ||v_{i,\mu_k}|| \leq \text{inj}(\mu_k).$$

The vectors $\left\{v_{i,\mu_k}\right\}_{i=1}^{m_1}$ are sorted according to their function values corresponding to the locally converted objective function $\tilde{f}_{\mu_k} = f \circ \exp_{\mu_k}$. The new search iterate (centroid) is thus obtained by mapping a weighted linear combination of $\left\{v_{i,\mu_k}\right\}_{i=1}^{m_1}$ from $T_{\mu_k}M$ to $M$ via the Riemannian exponential map, similar to Riemannian trust-region and Riemannian PSO:

$$\mu_{k+1} = \exp_{\mu_k}\left(\sum_{i=1}^{m_2} w_i v_i\right) \in N_{\mu_k},$$

**Algorithm 7.4:** CMA-ES on Euclidean space

**Data**: Initial mean $\mu_0 \in \mathbb{R}^n$, set initial covariance matrix $C$ the identity matrix $I$, step size $\sigma = 1$, $p_c = 0$, $p_\sigma = 0$.

**while** *stopping criterion not satisfied* **do**

Sample $\{v_i\}_{i=1}^{m_1} \sim N\left(\mathbf{0}, (\sigma)^2 \cdot C\right)$ ;

Update mean (search center) $\mu_{k+1} = \mu_k + \sum_{i=1}^{m_2} w_i v_i$ ;

Update $p_\sigma$ with $C, \sigma, m_{\text{eff}}, c_\sigma$ (isotropic evolution path):

$$p_\sigma = (1 - c_\sigma)\, p_\sigma + \sqrt{\left(1 - (1 - c_\sigma)^2\right)} \frac{\sqrt{m_{\text{eff}}}}{\sigma} \cdot C^{\frac{-1}{2}} \cdot \sum_{i=1}^{m_2} w_i v_i \ ;$$

Update $p_c$ with $\sigma, m_{\text{eff}}, c_c$ (anisotropic evolution path):

$$p_c = (1 - c_c)\, p_c + \sqrt{\left(1 - (1 - c_\sigma)^2\right)} \frac{\sqrt{m_{\text{eff}}}}{\sigma} \cdot \sum_{i=1}^{m_2} w_i v_i \ ;$$

Update covariance matrix $C$ with $\mu_{\text{cov}}, c_{\text{cov}}, p_\sigma, p_c$:

$$C = c_{\text{cov}} \cdot \left(1 - \frac{1}{\mu_{\text{cov}}}\right) \frac{1}{\sigma^2} \sum_{i=1}^{m_2} w_i v_i v_i^\top + (1 - c_{\text{cov}}) \cdot C + \frac{c_{\text{cov}}}{\mu_{\text{cov}}} p_c p_c^\top \ ;$$

Update step size $\sigma$ with $c_\sigma, d_\sigma, p_\sigma$:

$$\sigma = \sigma \cdot \exp\left(\frac{c_\sigma}{d_\sigma}\left(\frac{|p_\sigma|}{E|\,\mathrm{N}(0, I)\,|} - 1\right)\right);$$

k = k+1 ;

where $N_{\mu_k}$ is the normal neighbourhood of $\mu_k \in M$.

Astute readers may notice there is one caveat that differentiates the Riemannian version of the algorithm and its Euclidean counterpart: Tangent spaces are disjoint spaces, and the search information has to be carried from one search iterate to another in a way that is consistent with the manifold structure of the search spacc. Thc multi-dimensional search information such as covariance matrix $C$ and the evolutionary paths $p_\sigma$, $p_c$ are thus translated across search iterates with the notion of parallel transport. Note that all operations are done within a *single* normal neighbourhood, the parallel transport is described by Eq. (2.7).

Riemannian adaptation of CMA-ES is thus summarized by Algorithm 7.5 and the adaptation process described above are highlighted therein.

**Algorithm 7.5:** Riemannian CMA-ES

**Data**: Initial mean $\mu_0 \in \mathbb{R}^n$, set initial covariance matrix $C$ the identity matrix $I$, step size $\sigma = 1$, $p_c = 0$, $p_\sigma = 0$.

**while** *stopping criterion not satisfied* **do**

Sample $\{v_{i,\mu_k}\}_{i=1}^{m_1} \subset T_{\mu_k}M \sim N\left(\mathbf{0}, (\sigma)^2 \cdot C\right)$ ;

Update mean (search center) $\mu_{k+1} = \exp_{\mu_k}\left(\sum_{i=1}^{m_2} w_i v_i\right)$ ;

Update $p_\sigma$ with $C, \sigma, m_{\text{eff}}, c_\sigma$ (isotropic evolution path):

$$p_\sigma = (1 - c_\sigma)\, p_\sigma + \sqrt{\left(1 - (1 - c_\sigma)^2\right)} \frac{\sqrt{m_{\text{eff}}}}{\sigma} \cdot C^{\frac{-1}{2}} \cdot \sum_{i=1}^{m_2} w_i v_i \,;$$

Update $p_c$ with $\sigma, m_{\text{eff}}, c_c$ (anisotropic evolution path):

$$p_c = (1 - c_c)\, p_c + \sqrt{\left(1 - (1 - c_\sigma)^2\right)} \frac{\sqrt{m_{\text{eff}}}}{\sigma} \cdot \sum_{i=1}^{m_2} w_i v_i \,;$$

Update covariance matrix $C$ with $\mu_{\text{cov}}, c_{\text{cov}}, p_\sigma, p_c$:

$$C = c_{\text{cov}} \cdot \left(1 - \frac{1}{\mu_{\text{cov}}}\right) \frac{1}{\sigma^2} \sum_{i=1}^{m_2} w_i v_i v_i^\top + (1 - c_{\text{cov}}) \cdot C + \frac{c_{\text{cov}}}{\mu_{\text{cov}}} p_c p_c^\top \,;$$

Update step size $\sigma$ with $c_\sigma, d_\sigma, p_\sigma$:

$$\sigma = \sigma \cdot \exp\left(\frac{c_\sigma}{d_\sigma}\left(\frac{|p_\sigma|}{E|\,\mathrm{N}(0, I)\,|} - 1\right)\right)$$

Parallel transport $C, p_\sigma, p_c$ from $T_{\mu_k}M$ to $T_{\mu_{k+1}}M$ ; k = k+1 ;

## 7.3 Bridging Information Geometry, Stochastic Optimization and Riemannian Optimization

Recent developments in Information Geometry give raise to a geometrical interpretation of the decision space of population-based stochastic optimization algorithms on Euclidean spaces. Let $\mathcal{X} \subset \mathbb{R}^n$ denote the search space of an optimization problem. Let $S = \{p_\theta \mid \theta \in \Xi \subset \mathbb{R}^m\}$ denote a family of finitely parametrized probability densities *globally defined* over the search space $\mathcal{X}$. The statistical model $S$ constitutes the decision space of the optimization problem. By viewing $S$ as a statistical manifold (Chap. 3), the iterative evolution of statistical parameters in SDFO algorithms (described in Algorithm 7.3), can in some cases be thought of as a natural (Riemannian) gradient-based optimization process on $S$. Notable examples of natural gradient based optimization algorithms on $\mathcal{X} = \{0, 1\}^n$ [MMP11] and $\mathcal{X} = \mathbb{R}^n$ include Cross-Entropy [Rub97], (special cases of) CMA-ES [ANOK12, GSY+10] and Natural Evolution Strategies [WSPS08a], summarized by "Information Geometric Optimization" [OAAH17].

A natural question to ask is whether Riemannian adaptations of population-based stochastic optimization algorithms admit a similar Information Geometric interpretation. In particular, whether is it possible to view *both* the decision space and search space as Riemannian manifolds and relate the two geometrical structures.

In this section we first review natural-gradient based SDFO algorithms on Euclidean spaces. We then discuss how the locality of the Riemannian adaptation approach prevents a similar information geometric interpretation of Riemannian CMA-ES.

We begin by outlining SDFO algorithms under the paradigm of natural (Riemannian) gradient. Consider a **maximization** problem over Euclidean space $\mathcal{X} \subset \mathbb{R}^n$ defined by the objective function $f : \mathcal{X} \to \mathbb{R}$. Natural gradient algorithms can be summarized as follows (see also Algorithm 8.3 in the next chapter):

1. The optimization problem on $\mathcal{X}$ is transformed to an optimization problem on the decision space $\mathcal{S}$:

$$\max_{x \in \mathcal{X}} f(x) \mapsto \max_{p_\theta \in S} J_f(\theta) . \quad (7.4)$$

   This transformation arises in various forms, notable examples of this transformation include expected fitness [GSY+10, WSPS08b] (otherwise called stochastic relaxation [MMP11]) and quantile-based rewrite [OAAH17].
2. At each of the $k$th iteration for $k > 0$, the statistical parameter $\theta$ of $S$ is updated by the natural gradient with respect to the transformed objective function $J_f$ in Eq. (7.4):

$$\theta_{k+1} = \theta_k + \sigma_k \cdot \tilde{\nabla}_\theta J_f(\theta) ,$$

   where $\sigma_k$ denote the step size at the $k$th iteration, and $\tilde{\nabla}_\theta$ denote the natural gradient with respect to local coordinates $\theta$ and Fisher metric on $S$.

In this case, the decision space $S$ of natural gradient based algorithms is a statistical manifold, whereas the search space is usually a Euclidean space. The search trajectory can thus be understood as a curve on the statistical manifold decision space $S$, where the velocity is defined by the natural gradient with respect to the transformed objective function $J_f$. The above discussion is illustrated in Fig. 7.2.

A crucial assumption of all the natural gradient based algorithms is the existence of a family of ***globally defined*** finitely parametrized probability densities $S$ over the ***entire*** search space $\mathcal{X}$, whose statistical parameters coherent with the statistical estimations. However, as discussed in Chap. 4, this is generally not available on Riemannian manifolds.

Therefore, even though Euclidean SDFOs such as CMA-ES admit an information geometric interpretation [ANOK12], the Riemannian counterpart (described in [CFFS10] and Sect. 7.2.3) does not share the same property. In particular:

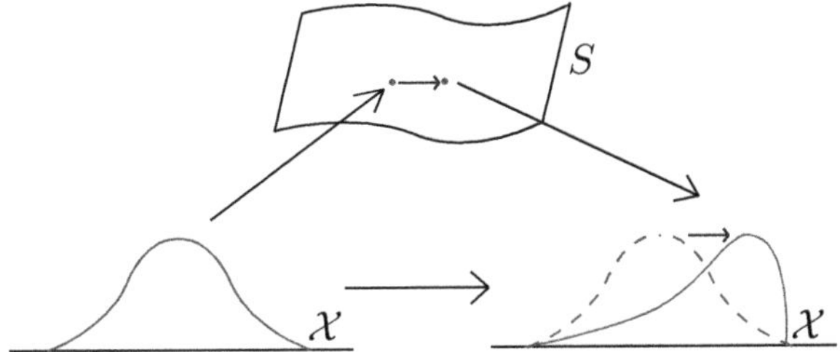


**Fig. 7.2** Illustration of an iteration of natural gradient based SDFOs. At each iteration, the algorithm begins with a sampling distribution on the search space $\mathcal{X}$ (bottom left). The evolution of sampling distributions (statistical parameter update), labeled by the blue arrow, is viewed as a natural gradient ascent/descent on an overarching statistical manifold decision space $S$ on the top picture. The new statistical parameter thus generates the new sampling distribution over the search space $\mathcal{X}$ (bottom right)

1. Riemannian CMA-ES adopts the "statistical" approach of locally inheriting probability distributions from tangent space centered at the search iterate to the manifold search space. The inherited density is supported within a *single* normal neighbourhood, hence the sampling and estimations are strictly local.
2. The tangent spaces $T_x M$ of a manifold $M$ are disjoint spaces, meaning locally inherited densities (Chap. 5, [Pen06]) supported in the normal neighbourhood of two search iterate/centroids are two *different* spaces.

In the next chapter, we first generalize Riemannian adaptation of SDFO and accentuate the aforementioned local restrictions. We then propose a population-based meta-algorithm by combining the information geometric view of SDFO and the geometrical framework described in Part I. The proposed algorithm both overcomes the local restrictions and provides a geometrical interpretation for stochastic optimization on Riemannian manifolds.

## References

[ABG07] P-A Absil, Christopher G Baker, and Kyle A Gallivan. Trust-region methods on Riemannian manifolds. *Foundations of Computational Mathematics*, 7(3):303–330, 2007.

[AH19] P-A Absil and S Hosseini. A collection of nonsmooth riemannian optimization problems. In *Nonsmooth Optimization and Its Applications*, pages 1–15. Springer, 2019.

[Ama98] Shun-Ichi Amari. Natural gradient works efficiently in learning. *Neural computation*, 10(2):251–276, 1998.

[AMS09] P-A Absil, Robert Mahony, and Rodolphe Sepulchre. *Optimization algorithms on matrix manifolds*. Princeton University Press, 2009.

[ANOK12] Youhei Akimoto, Yuichi Nagata, Isao Ono, and Shigenobu Kobayashi. Theoretical foundation for CMA-ES from information geometry perspective. *Algorithmica*, 64(4):698–716, 2012.

[Arn14] Dirk V Arnold. On the use of evolution strategies for optimization on spherical manifolds. In *International Conference on Parallel Problem Solving from Nature*, pages 882–891. Springer, 2014.

[BIA10] Pierre B Borckmans, Mariya Ishteva, and Pierre-Antoine Absil. A modified particle swarm optimization algorithm for the best low multilinear rank approximation of higher-order tensors. In *International Conference on Swarm Intelligence*, pages 13–23. Springer, 2010.

[CFFS10] Sebastian Colutto, Florian Fruhauf, Matthias Fuchs, and Otmar Scherzer. The cma-es on riemannian manifolds to reconstruct shapes in 3-d voxel images. *IEEE Transactions on Evolutionary Computation*, 14(2):227–245, 2010.

[EK95] Russell Eberhart and James Kennedy. Particle swarm optimization. In *Proceedings of the IEEE international conference on neural networks*, volume 4, pages 1942–1948. Citeseer, 1995.

[FMN16] Charles Fefferman, Sanjoy Mitter, and Hariharan Narayanan. Testing the manifold hypothesis. *Journal of the American Mathematical Society*, 29(4):983–1049, 2016.

[Gab82] Daniel Gabay. Minimizing a differentiable function over a differential manifold. *Journal of Optimization Theory and Applications*, 37(2):177–219, 1982.

[GSY+10] Tobias Glasmachers, Tom Schaul, Sun Yi, Daan Wierstra, and Jürgen Schmidhuber. Exponential natural evolution strategies. In *Proceedings of the 12th annual conference on Genetic and evolutionary computation*, pages 393–400. ACM, 2010.

[Han06] Nikolaus Hansen. The cma evolution strategy: a comparing review. In *Towards a new evolutionary computation*, pages 75–102. Springer, 2006.

[Han16] Nikolaus Hansen. The cma evolution strategy: A tutorial. *arXiv preprint* arXiv:1604.00772, 2016.

[HO96] Nikolaus Hansen and Andreas Ostermeier. Adapting arbitrary normal mutation distributions in evolution strategies: The covariance matrix adaptation. In *Evolutionary Computation, 1996., Proceedings of IEEE International Conference on*, pages 312–317. IEEE, 1996.

[KMH+04] Stefan Kern, Sibylle D Müller, Nikolaus Hansen, Dirk Büche, Jiri Ocenasek, and Petros Koumoutsakos. Learning probability distributions in continuous evolutionary algorithms–a comparative review. *Natural Computing*, 3(1):77–112, 2004.

[LL01] Pedro Larrañaga and Jose A Lozano. *Estimation of distribution algorithms: A new tool for evolutionary computation*, volume 2. Springer Science & Business Media, 2001.

[MMP11] Luigi Malagò, Matteo Matteucci, and Giovanni Pistone. Towards the geometry of estimation of distribution algorithms based on the exponential family. In *Proceedings of the 11th Workshop Proceedings on Foundations of Genetic Algorithms*, pages 230–242. ACM, 2011.

[NW06] Jorge Nocedal and Stephen Wright. *Numerical optimization*. Springer Science & Business Media, 2006.

[OAAH17] Yann Ollivier, Ludovic Arnold, Anne Auger, and Nikolaus Hansen. Information-geometric optimization algorithms: A unifying picture via invariance principles. *The Journal of Machine Learning Research*, 18(1):564–628, 2017.

[Pen06] Xavier Pennec. Intrinsic statistics on riemannian manifolds: Basic tools for geometric measurements. *Journal of Mathematical Imaging and Vision*, 25(1):127, 2006.

[Rub97] Reuven Y Rubinstein. Optimization of computer simulation models with rare events. *European Journal of Operational Research*, 99(1):89–112, 1997.

[WSPS08a] Daan Wierstra, Tom Schaul, Jan Peters, and Juergen Schmidhuber. Natural evolution strategies. In *2008 IEEE Congress on Evolutionary Computation (IEEE World Congress on Computational Intelligence)*, pages 3381–3387. IEEE, 2008.

[WSPS08b] Daan Wierstra, Tom Schaul, Jan Peters, and Jürgen Schmidhuber. Fitness expectation maximization. In *International Conference on Parallel Problem Solving from Nature*, pages 337–346. Springer, 2008.

# Chapter 8
# Stochastic Derivative-Free Optimization on Riemannian Manifolds

**Abstract** In this chapter, the main algorithm of the book, Extended RSDFO, is described. The chapter begins by formalizing a generalized framework, Riemannian Stochastic Derivative-Free Optimization (RSDFO) algorithms, for adapting Stochastic Derivative-Free Optimization (SDFO) algorithms from Euclidean spaces to Riemannian manifolds. RSDFO encompasses Riemannian adaptation of CMA-ES, and accentuates the main drawback of the Riemannian adaptation approach, i.e. the local restrictions. The chapter then describes Extended RSDFO, a principled population-based meta-algorithm that uses existing manifold optimization algorithms, such as RSDFO, as its local module. Extended RSDFO addresses the local restriction of RSDFO using both the intrinsic Riemannian geometry of the manifold $M$ (search space) and the product statistical Riemannian geometry of families of mixture densities over $M$ (decision spaces). The components of Extended RSDFO are constructed from a geometrical perspective, and its properties derived from first principles. In particular, we describe the geometry of the evolutionary steps of Extended RSDFO using a modified "regularized inverse" Fisher metric on the simplex of mixture coefficients, and derive the convergence behaviors of Extended RSDFO on compact connected Riemannian manifolds.

In this chapter we present Extended Riemannian Stochastic Derivative-Free Optimization (Extended RSDFO), a zero-order population-based stochastic meta-algorithm on Riemannian manifolds that overcomes the locality and implicit assumptions of manifold optimization in the literature.

Towards the end of the previous chapter, we discussed the locality of Riemannian adaptations of Euclidean optimization algorithms.This issue is particularly prevalent on Riemannian adaptation of Stochastic Derivative-Free Optimization (SDFO) algorithms such as Riemannian CMA-ES [CFFS10], where the local restrictions are imposed by *both* the Riemannian adaptation framework of optimization algorithms (Sect. 7.1) and the notion of locally inherited probability densities ([Pen06] and Chap. 5).



R. S. Fong and P. Tino, *Population-Based Optimization on Riemannian Manifolds*, Studies in Computational Intelligence 1046,
https://doi.org/10.1007/978-3-031-04293-5_8

Tangent spaces on manifolds are disjoint spaces, and families of locally inherited probability densities over them are *different* statistical manifolds. Whereas Euclidean spaces $\mathbb{R}^n$ are isomorphic to any of its tangent spaces, families of finitely parametrized probability densities defined *globally* on $\mathbb{R}^n$ can be regarded as a *single* statistical manifold. This means Riemannian adaptations of SDFO cannot be studied from the information geometric perspective as its Euclidean counterparts.

We begin our discussions by formalizing Riemannian Stochastic Derivative-Free Optimization (RSDFO) algorithms, a generalized framework for adapting SDFO algorithms from Euclidean spaces to Riemannian manifolds. RSDFO encompasses Riemannian adaptation of CMA-ES [CFFS10], and accentuates the main drawback, i.e. the local restrictions of the Riemannian adaptation approach. RSDFO also points us towards what is missing: finitely parametrized probability densities defined beyond the confines of a single normal neighbourhood on Riemannian manifolds, whose statistical parameters are coherent with statistical estimations. This is accomplished by the notion of mixture densities over totally bounded subsets on Riemannian manifolds described in Chap. 6.

We then describe Extended RSDFO, a principled population-based meta-algorithm that uses existing manifold optimization algorithms, such as RSDFO, as its local module. Extended RSDFO addresses the local restriction of RSDFO using both the intrinsic Riemannian geometry of the manifold $M$ (search space) and the product statistical Riemannian geometry of families of mixture densities over $M$ (decision spaces). The components of Extended RSDFO is constructed from a geometrical perspective, and its properties derived rigorously from first principles.

The remainder of the Chapter is outlined as follows:

1. In Sect. 8.1, we describe the Riemannian Stochastic Derivative-Free Optimization (RSDFO) algorithms, a generalized framework for adapting SDFO algorithms on Euclidean spaces to Riemannian manifolds. This serves as the core of Extended RSDFO and encompasses methods such as Riemannian adaptation of CMA-ES. We discuss the local restrictions of Riemannian adapted SDFO under the notion of locally inherited parametrized densities described in Chap. 5.
2. In Sect. 8.2, we describe the proposed meta-algorithm—Extended RSDFO. We begin by presenting an overview of Extended RSDFO. We then review the notion of parametrized mixture densities on totally bounded subsets of Riemannian manifolds (describe in Chap. 6) in the context of Extended RSDFO. In Sect. 8.2.2, we describe Extended RSDFO in algorithm form. The components of Extended RSDFO are detailed in the rest of the section.
3. In Sect. 8.3, we discuss the geometry of the evolutionary step of Extended RSDFO using a modified Fisher metric on the simplex of mixture coefficients. This provides us with a geometrical view of Extended RSDFO with Riemannian geometry of the search space and the Information Geometry of the decision space.
4. Finally in Sect. 8.4, we derive the convergence behaviors of Extended RSDFO. In particular, we show that Extended RSDFO converges in *finitely many* steps in compact connected Riemannian manifolds.

## 8.1 RSDFO: Riemannian Stochastic Derivative-Free Optimization Algorithms

In this section we describe a generalized framework—**Riemannian Stochastic Derivative-Free Optimization algorithms (RSDFO)**, for adapting Stochastic Derivative-Free Optimization (SDFO) algorithms from Euclidean spaces to Riemannian manifolds. This generalizes the approach of Riemannian adaptation of CMA-ES described in [CFFS10] using the notion of locally inherited parametrized probability distributions described in Chap. 5 and the Riemannian adaptation principle described in Chap. 7.

Let $(M, g)$ be a Riemannian manifold, and let the function $f : M \to \mathbb{R}$ denote the objective function of an optimization problem over $M$. For each $x \in M$, let $\tilde{f}_x := f \circ \exp_x : T_xM \to \mathbb{R}$ denote the locally converted objective function on each tangent space $T_xM$ of $M$. For each $x \in M$, let $U_x := \exp_x^{-1} N_x \subset T_xM$, centered at $\mathbf{0} \in T_xM$, denote the pre-image of a normal neighbourhood $N_x$ of $x$ under the Riemannian exponential map. We fix a predefined statistical model, denoted by $S_x := \left\{P\left(\cdot|\theta\right) \middle| \theta \in \Xi \subset \mathbb{R}^\ell, \ell \in \mathbb{N}_+\right\}$, throughout *all* pre-images of normal neighbourhoods in tangent spaces $U_x \subset T_xM$.

RSDFO is an iterative algorithm that generates search directions based on observations within the normal neighbourhood of the current search iterate $x \in M$. For each iteration, tangent vectors are sampled from a parametrized probability distribution $P\left(\cdot|\theta\right)$ in the predefined family $S_x$ over $U_x \subset T_xM$ centered at the search iterate $x \in M$. The fittest sampled tangent vectors are then used to determine the search direction and to estimate the statistical parameter $\theta \in \Xi$ for $S_x$. The new search iterate is obtained by tracing along the local geodesic starting at $x$ in the direction of the mean of the fittest tangent vectors, and the estimated statistical parameters of $S_x$ are subsequently parallel transported to the new search iterate. This procedure is summarized in Algorithm 8.1.[1]

**Algorithm 8.1:** Riemannian SDFO

**Data**: Initial point $\mu_0 \in M$, initial statistical parameter $\theta^0 \in \Xi$, set iteration counter $k = 0$, and step size functions $\sigma_\theta(k)$, $\sigma_v(k)$.

**while** *stopping criterion not satisfied* **do**
- Generate set of sample vectors $V^k \sim P(\cdot|\theta^k) \in S_{\mu_k} \subset \text{Prob}(U_{\mu_k})$ ;
- Evaluate locally converted fitness $\tilde{f}_{\mu_k}(v)$ of each $v \in V^k \subset T_{\mu_k}M$ ;
- Estimate statistical parameter $\hat{\theta}^{k+1}$ based on the fittest tangent vectors $\hat{V}^k \subset V^k$ ;
- Translate the center of search: $\hat{v}_k := \sigma_v(k) \cdot \mu(\hat{V}^k)$, and $\mu_{k+1} := \exp_{\mu_k}(\hat{v}_k)$.
- Parallel transport the statistical parameters: $\theta^{k+1} := P_k^{k+1}\left(\sigma_\theta(k) \cdot \hat{\theta}^{k+1}\right)$, where $P_k^{k+1} : T_{\mu_k}M \to T_{\mu_{k+1}}M$ denote the parallel transport from $T_{\mu_k}M$ to $T_{\mu_{k+1}}M$ ;
- k = k+1;

[1] Note that $V^k \subset U_{\mu_k} \subset T_{\mu_k}M$ are vectors on the tangent space, which is a Euclidean space.

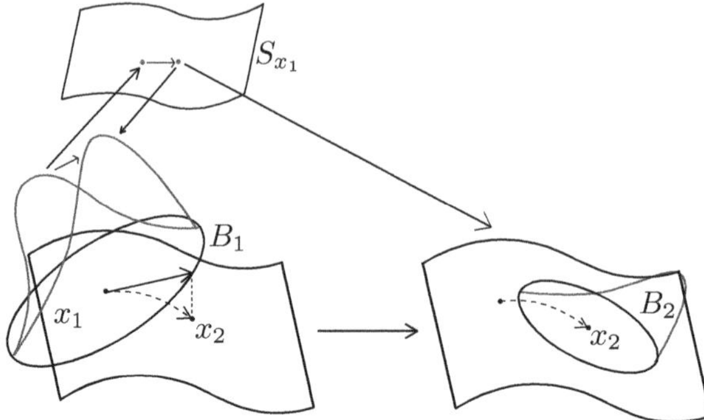


**Fig. 8.1** Illustration of an iteration of Riemannian adapted natural-gradient based SDFOs. For each iteration, the algorithm initiates with $x_1 \in M$. A set of tangent vectors is drawn from a sampling distribution on $B_1 \subset T_{x_1}M$, and a new sampling distribution is estimated by natural-gradient on the overarching statistical manifold $S_{x_1}$. The new sampling distribution is then parallel transported to the new search iterate $x_2 \in M$. Notice the estimations and computations are performed in a *strictly local* fashion within the tangent spaces of the search iterates

Riemannian adapted natural-gradient based SDFOs admit a *strictly local* information geometric interpretation on the Riemannian manifold search space, contrary to the *global* information geometric view of their Euclidean counterparts described in Sect. 7.3.This is illustrated in Fig. 8.1 by combining Figs. 7.1 and 7.2.

### *8.1.1 Discussion, Shortcoming and Improvements*

In RSDFO, the parametrized densities are inherited from tangent spaces to manifolds locally via Riemannian exponential map [Pen04] or orientation-preserving diffeomorphisms (Chap. 5). As tangent spaces are vector spaces, the local computations can be perform on the tangent spaces in the same fashion as the pre-adapted algorithm. This means RSDFO preserves both the local step-wise computations and the overall structure of the original pre-adapted Euclidean SDFO algorithm. The only additional tool needed to translate the search information from the tangent spaces back to the Riemannian manifold was the Riemannian exponential map (or orientation-preserving diffeomorphisms) and to some extent the knowledge of the injectivity radius of the Riemannian manifold.

However, Riemannian SDFO algorithms constructed under the Riemannian adaptation approach also has two drawbacks, which are imposed by the locality of normal neighbourhoods and the notion of locally inherited probability densities:

- First of all, the locality of the inherited parametrized densities implies the quality of the Riemannian version of SDFOs depends strongly on the structure of the manifold. For each iteration, the local search region centered at $x \in M$ is bounded by the normal neighbourhood $N_x$ of $x$. Manifolds with high scalar curvature has small

injectivity radius, and thus small normal neighbourhoods [AM97, Kli61, CEE75]. In this case the local observations, bounded by the normal neighbourhoods, are highly restrictive.
Whilst SDFO such as CMA-ES have an advantage in tackling multi-model optimization problems [HK04], this advantage may not be reflected on the Riemannian counterparts on more complex manifolds.

- Secondly, tangent spaces of a manifold are disjoint spaces. The local search regions of RSDFO, given by the normal neighbourhoods centered at the search iterates, varies from iteration to iteration. The families of locally inherited densities supported on different local search regions are therefore *different* spaces, even if they share the same parametrization. Statistical estimations such as empirical moments around the search iterates therefore cannot be directly compared. Even though convergence behaviours of Euclidean SDFO have been studied in [ZM04, Bey14], the Riemannian counterpart cannot be analysed in a similar fashion as there is no explicit relation between families of sampling densities across iterations.

In order to overcome the local restrictions of the RSDFO under the Riemannian adaptation approach, we require a family of intrinsic, computable and parametrized distributions defined beyond the confines of a single normal neighbourhoods on Riemannian manifolds. This is accomplished by the notion of mixture densities described in Chap. 6. In the remainder of the chapter we will address the local restrictions of generic RSDFO algorithms with a zero-order population-based stochastic meta-algorithm—Extended RSDFO.

## 8.2 Extended RSDFO on Riemannian Manifolds

We begin our discussion by an overview of **Extended RSDFO**. For each iteration of RSDFO (Algorithm 8.1), the search information is generated locally within the normal neighbourhood of each search iterate. This means at each iteration, the search region of the algorithm is confined by the normal neighbourhood centered at the search iterate.

In order to overcome this local restriction, Extended RSDFO considers *multiple* search centroids simultaneously instead of a *single* search iterate. Moreover, we optimize over a set of overlapping geodesic balls covering the manifold centering around each of the centroids, instead of the centroids themselves. This process is guided by families of parametric mixture densities over the manifold, allowing us to relate the families of probability distributions supported on the different tangent spaces of the set of search centroids.

Fixing an RSDFO algorithm as the core local module, a single step of Extended RSDFO is summarized as follows: at each iteration, Extended RSDFO initiates with a finite set of centroids on the Riemannian manifold. The search region of the iteration is thus given by the union of the set of geodesic balls of injectivity radius centered at the centroids, forming a totally bounded subset of the search space manifold. In

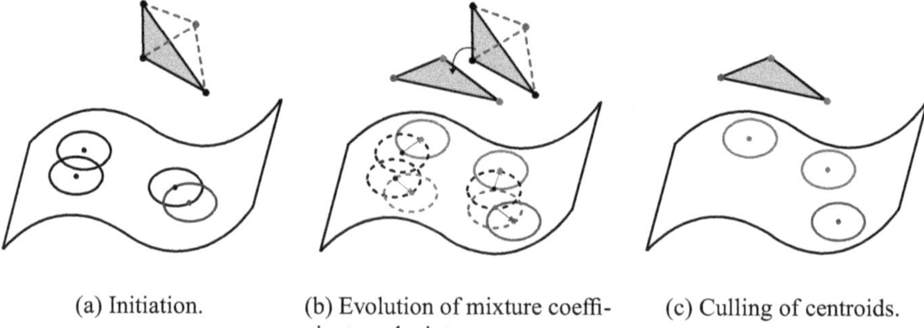

(a) Initiation.

(b) Evolution of mixture coefficients and mixture components.

(c) Culling of centroids.

**Fig. 8.2** Illustration (from left to right) of the "full" evolutionary step of Extended RSDFO. At each iteration, Extended RSDFO begins with a set of centroids (black) on the manifold (Fig. 8.2a). An additional centroid (red), is generated on the boundary of the explored region. In Fig. 8.2b, each centroid generates a new "offspring" with a step of RSDFO core (green). The centroids are then culled according to their relative expected fitness. The culled centroids are marked by the dotted lines. The remaining centroids move onto the next iteration, as illustrated on the right image (Fig. 8.2c). The evolution of mixture coefficients corresponding to the search centroids is represented by the change of shaded region on the simplex above the search space manifold (Sect. 8.3)

addition, we also generate another set of "exploration" centroids on the boundary of explored region.

We then perform one iteration of the predefined RSDFO algorithm on a subset of the generated centroids, which then generates a new set of search centroids (and the corresponding geodesic balls of injectivity radius) on the manifold. The sets of all search centroids are then compared based on the relative expected fitness over the geodesic balls centered around them, and the fittest centroids are carried over to the next iteration. Each "stream" of local RSDFO search shares a parametrized family of locally inherited densities, whereas a boundary exploration point initiates a new "stream". The above discussion is summarized in Fig. 8.2.[2]

The objective of the Extended RSDFO is to find a *set* of fittest centroids (and their corresponding geodesic balls) on the search space manifold. The weight of each centroid is determined by the mixture coefficients of the overarching mixture densities, which are supported on the union of geodesic balls centered at the set of centroids.

Furthermore, as the overarching family of mixture densities over the union of geodesic balls relates the local families of probability densities support on each of the local geodesic balls. The evolutionary step of Extended RSDFO can therefore be

[2] It is worth noting that for this illustration (Fig. 8.2), the full interim simplex in the middle image consists of 8 vertices, where each vertex corresponds to the mixture coefficient of each of the 8 interim centroids (initial, generated, boundary). The "original" simplex (left picture) and the "target" simplex share at most $(3 + 1 + 4) - 4 - 1 = 3$ number of vertices (in this case none). For each step of the algorithm we transverse through a different interim simplex in this fashion. One can picture the evolutionary step of the algorithm in the context of simplex of mixture coefficients as jumping from one subsimplex to another.

viewed as a change of mixture densities over the "accumulative" union of geodesic balls over $M$, guided by the change of face in the simplex of mixture coefficients. This enables us to both study the evolution step and analyse the convergence behaviours of Extended RSDFO from a geometrical perspective.

For each iteration, a typical search region of Extended RSDFO is given by a union of geodesic balls, forming a totally bounded subset in the search space manifold $M$. In order to establish the necessary tools to formally describe the algorithm, we take a slight detour in the next subsection by reviewing the notion of parametrized mixture distributions on *union of geodesic balls* of Riemannian manifolds. The detailed construction and dualistic geometry [AN00] of the mixture densities on *totally bounded subsets* of Riemannian manifolds is detailed in the first part of the book.

### 8.2.1 Parametrized Mixture Distribution on Totally Bounded Subsets of Riemannian Manifold

In this section we recall from Chap. 6 the notion of finitely parametrized mixture densities in the context of union of geodesic balls, which forms a totally bounded subset of the search space manifold. Let $(M, g)$ be an smooth $n$-dimensional Riemannian manifold, an orientation-preserving open cover of $M$ is an at most countable set of pairs $\{(\rho_\alpha, U_\alpha)\}_{\alpha\in\Lambda_M}$ satisfying:

1. $U_\alpha \subset \mathbb{R}^n$, $\rho_\alpha : U_\alpha \to M$ are orientation preserving diffeomorphisms for each $\alpha \in \Lambda_M$, and
2. the set $\{\rho_\alpha(U_\alpha)\}_{\alpha\in\Lambda_M}$ is an open cover of $M$.

For each $x_\alpha \in M$, let $j_{x_\alpha} \leq \text{inj}(x_\alpha)$ and let $\mathcal{E}_M := \left\{\left(\exp_{x_\alpha}, B(\mathbf{0}, j_{x_\alpha})\right)\right\}_{\alpha\in\Lambda_M}$ denote an orientation-preserving open cover of $M$ by geodesic balls indexed by an at most countable[3] set $\Lambda_M$. Given a finite set of centroids $\{x_\alpha\}_{\alpha\in\Lambda_V}$, indexed by a finite subset $\Lambda_V \subset \Lambda_M$, the local search region at an iteration of Extended RSDFO is given by a union of geodesic balls:

$$V := \bigcup_{\alpha\in\Lambda_V} \exp_{x_\alpha}\left(B(\mathbf{0}, j_{x_\alpha})\right) \subset M \quad . \tag{8.1}$$

Let $\mathcal{E}_V := \left\{\left(\exp_{x_\alpha}, B(\mathbf{0}, j_{x_\alpha})\right)\right\}_{\alpha\in\Lambda_V} \subset \mathcal{E}_M$, denote the finite sub-cover of $\mathcal{E}_M$ over $V$. For each $\alpha \in \Lambda_V$, let $m_\alpha \in \mathbb{N}_+$, and let $S_\alpha := \{p(\cdot|\theta^\alpha)|\theta^\alpha \in \Xi_\alpha \subset \mathbb{R}^{m_\alpha}\} \subset \text{Prob}(B(\mathbf{0}, j_{x_\alpha}))$ denote a family of finitely parametrized probability densities over $B(\mathbf{0}, j_{x_\alpha}) \subset T_{x_\alpha}M$.

By the discussion in Chap. 5, we can define for each $\alpha \in \Lambda_V$ a family of locally inherited probability densities on *each* geodesic ball $\exp_{x_\alpha}\left(B(\mathbf{0}, j_{x_\alpha})\right) \subset M$:

[3] This is due to the fact that manifolds are second-countable, thus Lindelof, meaning all open cover has an at most countable sub-cover.

$$\tilde{S}_\alpha := \exp_\alpha^{-1^*} S_\alpha = \left\{ \tilde{p}(\cdot|\theta^\alpha) = \exp_\alpha^{-1^*} p(\cdot|\theta^\alpha) \right\} \quad .$$

Consider the closure of the simplex of mixture coefficients denoted by:

$$\overline{S}_0 := \left\{ \{\varphi_\alpha\}_{\alpha\in\Lambda_V} \,\middle|\, \varphi_\alpha \in [0,1]\,,\ \sum_{\alpha\in\Lambda_V} \varphi_\alpha = 1 \right\} \subset [0,1]^{|\Lambda_V|-1} \quad . \tag{8.2}$$

A family of parametrized mixture densities on $V$ can thus be defined by:

$$\mathcal{L}_V := \left\{ \sum_{\alpha\in\Lambda_V} \varphi_\alpha \cdot \tilde{p}(\cdot|\theta^\alpha) \,\middle|\, \{\varphi_\alpha\}_{\alpha\in\Lambda_V} \in \overline{S}_0,\ \tilde{p}(\cdot|\theta^\alpha) \in \tilde{S}_\alpha \right\} \quad . \tag{8.3}$$

By the discussion in Sect. 6.4, the set of mixture densities $\mathcal{L}_V$ is a product Riemannian manifold given by: $\mathcal{L}_V = \overline{S}_0 \times \bigoplus_{\alpha\in\Lambda_V} \tilde{S}_\alpha$.[4]

### 8.2.2 Extended RSDFO

In this section we describe the Extended RSDFO algorithm. Let $(M, g)$ be a Riemannian manifold search space, and let $f : M \to \mathbb{R}$ denote the real-valued objective function of a *maximization* problem over $M$. We assume, without loss of generality, that $f$ is a strictly positive function, i.e. $f(x) > 0$ for all $x \in M$. Analogously we assume $f$ is strictly negative for minimization problems, the minimization case is discussed in Remark 8.3.

Given a point $x_\alpha \in M$, the **expected fitness** [WSPS08b](or **stochastic relaxation** in [MP15]) of $f$ over probability density $p(x|\theta^\alpha)$ locally supported in geodesic ball $\exp_{x_\alpha}\left(B(\mathbf{0}, j_{x_\alpha})\right)$ is the real number:

$$E_\alpha := \int_{\exp_{x_\alpha}(B(0, j_{x_\alpha}))} f(x)\tilde{p}(x|\theta^\alpha)dx \quad . \tag{8.4}$$

In practice, this integral can be approximated using $\tilde{p}(x|\theta^\alpha)$−Monte Carlo samples for $x \in M$. Note that this is an integral over the Riemannian manifold $M$, for further detail see [Lee01].

Choose and fix an RSDFO (Algorithm 8.1) method as the core local module, the Extended RSDFO on Riemanian manifolds is summarized in Algorithm 8.2.

Note that in Line 3 of Extended RSDFO (Algorithm 8.2): each search centroid has its own parametrized family of locally inherited probability densities. The statistical parameters on the new centroids are parallel translated from their previous iterate, as described in Sect. 2.3.1 and Algorithm 8.1 in Sect. 7.1.

---

[4] The conditions required in Sect. 6.3.1 are naturally satisfied by the construction of $V$ and the families of inherited densities described in this paper.

**Algorithm 8.2:** Extended RSDFO

**Data**: Initial set of centroids in $M$: $X^0 := \{x_\alpha\}_{\alpha \in \Lambda^0}$, $\Lambda^0 \subset \Lambda_M$, initial set of parameters for the mixture distribution: $\{\varphi_\alpha^0, \theta_0^\alpha\}_{\alpha \in \Lambda^0}$. Set iteration count $k = 0$, choose non-increasing non-zero function $0 < \tau(k) \leq 1$. Integers $N_{cull}, N_{random} > 0$.

1 **while** *stopping criterion not satisfied* **do**

2 Generate $N_{random}$ non-repeating centroids $\overline{X}^{k+1}$ from the mixture distribution:

$$P_k(\cdot) = (1 - \tau(k)) \cdot \sum_{\alpha \in \Lambda^k} \varphi_\alpha^k \delta_{x_\alpha} + \tau(k) \cdot U \quad ,$$

where $\delta_{x_\alpha}$ is the delta function on $x_\alpha \in X^k$, and $U$ denote the exploration distribution described in Sect. 8.2.6 below ;

3 **Evolution of mixture components:** For each $x_\alpha \in \overline{X}^{k+1}$ perform one step of the chosen RSDFO, generate new set of centroids $\hat{X}^{k+1}$, and the corresponding statistical parameter $\theta_{k+1}^\alpha$ for each $x_\alpha \in \hat{X}^{k+1}$. ;

4 We now have interim centroids $X^k \cup \hat{X}^{k+1}$, indexed by $\hat{\Lambda}^{k+1}$, where $|\hat{\Lambda}^{k+1}| = |X^k \cup \hat{X}^{k+1}| < \infty$. ;

5 **Evolution of mixture coefficients:** Evaluate the expected fitness $E_\alpha$ for each $x_\alpha \in X^k \cup \hat{X}^{k+1}$. ;

6 Preserve the fittest $N_{cull}$ centroids $X^{k+1} := \{x_\alpha\}_{\alpha \in \Lambda^{k+1}}$, indexed by $\Lambda^{k+1} \subset \hat{\Lambda}^{k+1}$. This defines the search region of the next iteration. ;

7 Recalibration of mixture coefficients $\{\varphi_\alpha^{k+1}\}_{\alpha \in \Lambda^{k+1}}$ described in Eq. (8.10) in Sect. 8.2.4 ;

8 k = k+1;

### 8.2.3 Additional Parameters

In addition to the native parameters inherited from the pre-defined RSDFO algorithm, Extended RSDFO requires *three* additional parameters: two positive integers $N_{random}, N_{cull} \in \mathbb{N}_+$ to control the number of search centroids during and after the iteration respectively, and a non-increasing, non-zero function $\tau(k)$ on the iteration counter $k$. The function $\tau(k)$ decays to an arbitrary small positive constant $\epsilon_c \in (0, 1)$, such that as $k \to \infty$: $1 \gg \epsilon_c > 0$ . That is, the sampling centroids will eventually be mostly from the distribution $\sum_{\alpha \in \Lambda^k} \varphi_\alpha^k \delta_{x_\alpha}$.

### 8.2.4 Evolutionary Step

In this section, we discuss the evolutionary steps of Algorithm 8.2 in the context of family of mixture distributions described in Sect. 8.2.1 (illustrated in Fig. 8.2). The geometrical interpretations and justifications will be fleshed out in Sect. 8.3.

In the $k$th iteration for $k > 0$, a finite set of centroids $X^k := \{x_\alpha\}_{\alpha \in \Lambda^k}$ is inherited from the previous iteration on the Riemannian manifold, and a new set of centroids $\hat{X}^{k+1}$ is generated through lines 2 and 3 of Extended RSDFO (Algorithm 8.2). The union of the two sets of centroids: $X^k \cup \hat{X}^{k+1}$, indexed by $\hat{\Lambda}^{k+1}$, forms the centroids of the interim search region of the $k$th iteration.

The set of interim centroids $X^k \cup \hat{X}^{k+1}$ is subsequently culled according to their expected fitness, and the fittest centroids $X^{k+1} := \{x_\alpha\}_{\alpha \in \Lambda^{k+1}}$, indexed by $\Lambda^{k+1} \subset \hat{\Lambda}^{k+1}$, are preserved for the next iteration. This process is repeated until a termination criteria is satisfied (see Sect. 8.2.7).

More formally, let $\left\{\exp_{x_\alpha}\left(B(\mathbf{0}, j_{x_\alpha})\right)\right\}_{\alpha \in \hat{\Lambda}^{k+1}}$ denote the set of closed geodesic ball of injectivity radius centered at $\{x_\alpha\}_{\hat{\Lambda}^{k+1}} := X^k \cup \hat{X}^{k+1}$. The region in $M$ explored by the $k$th iteration of Algorithm 8.2 can be separated into three parts:

$$V^k := \bigcup_{\alpha \in \Lambda^k} \tilde{B}_\alpha \ , \quad V^{k+1} := \bigcup_{\alpha \in \Lambda^{k+1}} \tilde{B}_\alpha, \quad \text{and} \tag{8.5}$$
$$\hat{V}^{k+1} := \bigcup_{\alpha \in \hat{\Lambda}^{k+1}} \tilde{B}_\alpha \supseteq V^k \cup V^{k+1} \ ,$$

where $\tilde{B}_\alpha := \exp_{x_\alpha}\left(B(\mathbf{0}, j_{x_\alpha})\right) \subset M$ for simplicity. We begin the $k$th iteration with the search region $V^k$, which expands to the interim search region $\hat{V}^{k+1}$. The interim search region $\hat{V}^{k+1}$ is then reduced to $V^{k+1} \subset \hat{V}^{k+1}$, the search region for the $(k+1)$th iteration, after the culling of centroids.

As a result, for each iteration of Extended RSDFO, we would transverse through *three* families of parametrized mixture densities supported on the three subsets $V^k, V^{k+1}, \hat{V}^{k+1} \subset M$ respectively. The families of parametrized mixture densities are given as follows, a summary of the notation can be found in Table 8.1.

First consider the closure of simplices of mixture coefficients of each $\mathcal{L}_{V^k}, \mathcal{L}_{V^{k+1}}, \mathcal{L}_{\hat{V}^{k+1}}$ given by:

$$\overline{S}_0^k := \left\{ \{\varphi_\alpha\}_{\alpha \in \Lambda^k} \,\middle|\, \varphi_\alpha \in [0, 1], \sum_{\alpha \in \Lambda^k} \varphi_\alpha = 1 \right\} \tag{8.6}$$

$$\hat{S}_0^{k+1} := \left\{ \{\varphi_\alpha\}_{\alpha \in \hat{\Lambda}^{k+1}} \,\middle|\, \varphi_\alpha \in [0, 1], \sum_{\alpha \in \hat{\Lambda}^{k+1}} \varphi_\alpha = 1 \right\} \tag{8.7}$$

$$\overline{S}_0^{k+1} := \left\{ \{\varphi_\alpha\}_{\alpha \in \Lambda^{k+1}} \,\middle|\, \varphi_\alpha \in [0, 1], \sum_{\alpha \in \Lambda^{k+1}} \varphi_\alpha = 1 \right\} \ . \tag{8.8}$$

**Table 8.1** Summary of notations in the evolutionary step of the $k$th iteration of Extended RSDFO

| Mixture family | Simplex | Index | Search region | Centroids | Description |
|---|---|---|---|---|---|
| $\mathcal{L}_{V^k}$ | $\overline{S}_0^k$ | $\Lambda^k$ | $V^k$ | $X^k$ | Generated by centroids from the $(k-1)$th iteration |
| $\mathcal{L}_{V^{k+1}}$ | $\overline{S}_0^{k+1}$ | $\Lambda^{k+1}$ | $V^{k+1}$ | $X^{k+1}$ | Generated by new centroids for the $(k+1)$th iteration from the $k$th iteration |
| $\mathcal{L}_{\hat{V}^{k+1}}$ | $\hat{S}_0^{k+1}$ | $\hat{\Lambda}^{k+1}$ | $\hat{V}^{k+1}$ | $X^k \cup \hat{X}^{k+1}$ | Generated by the interim centroids within the $k$th iteration. $\hat{S}_0^{k+1} \supset \overline{S}_0^{k+1}, \overline{S}_0^k$ |

Note that the simplices $\overline{S}_0^k$, $\overline{S}_0^{k+1}$ are in fact two faces of $\hat{S}_0^{k+1}$ sharing *at most*[5] $\left(|\Lambda^k| + N_{random}\right) - N_{random} = (N_{cull} + N_{random}) - N_{random} = N_{cull}$ vertices. An example of the simplices is illustrated in Fig. 8.3.

The three mixture families in the $k$th iteration, summarized in Table 8.1, are thus given by:

$$\mathcal{L}_V = \left\{ \sum_{\alpha \in \Lambda_V} \varphi_\alpha \cdot \tilde{p}(\cdot|\theta^\alpha) \middle| \{\varphi_\alpha\}_{\alpha \in \Lambda_V} \in \overline{S}_0,\ \tilde{p}(\cdot|\theta^\alpha) \in \tilde{S}_\alpha \right\} , \tag{8.9}$$

where $V = V^k, \hat{V}^{k+1}, V^{k+1} \subset M$ denoting the searching region generated by the set of centroids $X = X^k, X^k \cup \hat{X}^{k+1}, X^{k+1} \subset M$ respectively described in Eq. (8.5). The sets of indices of $X = X^k, X^k \cup \hat{X}^{k+1}, X^{k+1}$ are denoted by $\Lambda_V = \Lambda^k, \hat{\Lambda}^{k+1}, \Lambda^{k+1}$ respectively, and the corresponding closure of simplices of mixture coefficients is given by $\overline{S}_0 = \overline{S}_0^k, \hat{S}_0^{k+1}, \overline{S}_0^{k+1}$. For all $\alpha \in \Lambda_V$, where $\Lambda_V = \Lambda^k, \hat{\Lambda}^{k+1}, \Lambda^{k+1}$, the set $\tilde{S}_\alpha := \exp_\alpha^{-1^*} S_\alpha$ are locally inherited parametrized densities on $\tilde{B}_\alpha \subset M$ described in Chap. 5. A summary of the notations can be found in Table 8.1. Detailed descriptions of the geometry of $\mathcal{L}_V$ is in Chap. 6.

Due to the product manifold structure of the family of mixture densities $\mathcal{L}_V = \overline{S}_0 \times \bigoplus_{\alpha \in \Lambda_V} \tilde{S}_\alpha$, the evolution of statistical parameters of the mixture component densities $\tilde{S}_\alpha$ for $\alpha \in \Lambda_V$ and the evolution of mixture coefficients in $\overline{S}_0$ in Extended

[5] This can be zero, as illustrated in Fig. 8.2.

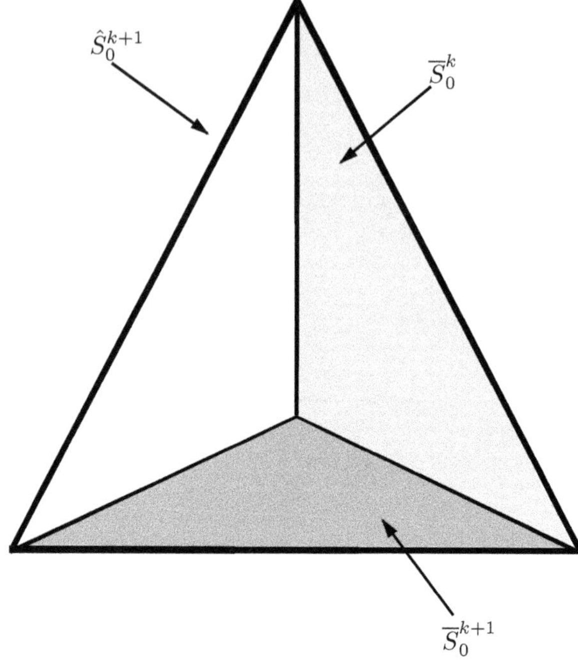


**Fig. 8.3** Simplicial illustration of the simplices in evolutionary step of Extended RSDFO (Algorithm 8.2). The interim simplex $\hat{S}_0^{k+1}$ is given by the entire simplex with 4 vertices. The subsimplices $\overline{S}_0^{k}$ and $\overline{S}_0^{k+1}$ is given by the right and bottom simplices (shaded) with 3 vertices respectively

RSDFO can be handled *separately* and *independently*. As the evolution of component densities in $\tilde{S}_\alpha$ are already described by the pre-defined RSDFO in Algorithm 8.1 for each $\alpha \in \Lambda_V$, it remains to describe the evolution of mixture coefficients.

Given that for each center indexed by $\alpha$ we quantify the quality of solutions around it by the expected fitness $E_\alpha$, it is natural to assign to the individual mixture coefficients $\varphi_\alpha$ a value proportional to $E_\alpha$. In particular, in line 7 of Algorithm 8.2 the mixture coefficients are updated via the following equation:

$$\varphi_\alpha^{k+1} = \frac{E_\alpha}{\sum_{\alpha \in \Lambda^{k+1}} E_\alpha} \ . \tag{8.10}$$

where $E_\alpha$ is the expected fitness of $\tilde{B}_\alpha$ described in line 5 in Algorithm 8.2 above. Note that due to the assumption that $f$ is strictly positive, the expected fitness $E_\alpha$ is also strictly positive for all $\alpha$. A detailed theoretical justification will be discussed in Sect. 8.3. The case where we have a minimization problem is discussed in Remark 8.3, in particular in Eq. (8.27).

### 8.2.5 Monotone Improvement on Expected Fitness

Recall the discussion in Sect. 8.1.1: given two search centroids (or search iterates) $x_k, x_{k+1} \in M$ in Riemannian adapted optimization algorithms, the tangent spaces $T_{x_k}M$ and $T_{x_{k+1}}M$ centered at $x_k, x_{k+1}$ respectively are *different* (disjoint) spaces. In the case of RSDFO, the locally inherited parametrized probability densities (see Chap. 5) over the tangent spaces $T_{x_k}M$ and $T_{x_{k+1}}M$ have *different* supports, thus they belong to different statistical manifolds / models. In classical RSDFO, the search information therefore cannot be compared beyond the objective value of the search iterates.

On the other hand in Extended RSDFO, the family of mixture densities $\mathcal{L}_V$ relates the sets of parametrized densities across different tangent spaces. In particular, on the interim search region $\hat{V}^{k+1}$ corresponding to the interim search centroids $X^k \cup \hat{X}^{k+1}$, the sampling distributions of the $k$th and $(k+1)$th can be regarded as distributions on the *same* family of mixture densities $\mathcal{L}_{\hat{V}^{k+1}}$. Therefore it is reasonable to compare their expected fitness on the interim search region $\hat{V}^{k+1}$. This allows us to derive the following result.

**Proposition 8.1** *The sequence of expected fitness across iterations of Extended RSDFO on a maximization problem is monotonic, non-decreasing. That is, for each iteration $k > 0$, let $X^k = \{x_\alpha\}_{\alpha \in \Lambda^k}$ denote the search centroids of initial search region $V^k$ indexed by $\Lambda^k$. Let $\left\{\varphi_\alpha^k\right\}_{\alpha \in \Lambda^k}$ denote the set of mixture coefficients of $\left\{\exp_{x_\alpha}\left(B(\mathbf{0}, j_{x_\alpha})\right)\right\}_{\alpha \in \Lambda^k}$ described in Eq.* (8.10)*, then the following is satisfied:*

$$E^k := \sum_{\alpha \in \Lambda^k} \varphi_\alpha^k E_\alpha^k \leq \sum_{\alpha \in \Lambda^{k+1}} \varphi_\alpha^{k+1} E_\alpha^{k+1} =: E^{k+1} \quad ,$$

*where $E_\alpha^k$ denote the expected fitness of $\exp_{x_\alpha}\left(B(\mathbf{0}, j_{x_\alpha})\right)$ for $x_\alpha \in X^k$, and $E^k$ denote the expected fitness of the sampling distribution at the* $k$th *iteration of Extended RSDFO.*

***Proof*** Let $f : M \to \mathbb{R}$ be the objective function of a maximization problem over $M$. At the $k$th iteration of Extended RSDFO, let $X^k = \{x_\alpha\}_{\alpha \in \Lambda^k}$ denote the search centroids of search region $V^k$ indexed by $\Lambda^k$. Let $\mathcal{L}_{V^k} = \overline{S}_0^k \times \bigoplus_{\alpha \in \Lambda^k} \tilde{S}_\alpha$ be the family of mixture densities on $X^k$, described in Eq. (8.3), with simplex of mixture coefficients $\overline{S}_0^k$ and family of locally inherited mixture component densities $\tilde{S}_\alpha$. Let $\xi^k := \left\{\left\{\varphi_\alpha^k\right\}_{\alpha \in \Lambda_V}, \left\{\theta_k^\alpha\right\}_{\alpha \in \Lambda^k}\right\}$ denote the set of parameters of the product statistical manifold $\mathcal{L}_{V^k} = \overline{S}_0^k \times \bigoplus_{\alpha \in \Lambda^k} \tilde{S}_\alpha$.

For each $p(x|\xi^k) := \sum_{\alpha \in \Lambda^k} \varphi_\alpha^k \cdot \tilde{p}(x|\theta_k^\alpha) \in \mathcal{L}_{V^k} \subset \text{Prob}(V)$, the expected fitness (Eq. (8.4)) of $f$ over $p(x|\xi^k)$ (supported in $V^k$) on Riemannian manifold search space $M$ is given by:

$$
\begin{aligned}
E^k &:= \int_{x\in V^k} f(x)p(x|\xi^k)dx \\
&= \sum_{\alpha\in\Lambda^k} \varphi_\alpha^k \cdot \int_{x\in\tilde{B}_\alpha} f(x)\tilde{p}(x|\theta_k^\alpha)dx \quad ,
\end{aligned}
\tag{8.11}
$$

where $\tilde{B}_\alpha := \exp_{x_\alpha}\left(B(\mathbf{0}, j_{x_\alpha})\right)$ and $\tilde{p}(x|\theta_k^\alpha) \in \tilde{S}_\alpha := \exp_{x_\alpha}^{-1^*} S_\alpha$ are locally inherited parametrized probability densities described in Chap. 5.

Let $X^k \cup \hat{X}^{k+1}$ denote the interim centroids of the $k$th iteration, indexed by $\hat{\Lambda}^{k+1}$, corresponding to the interim search region $\hat{V}^{k+1} \subset M$ described in Eq. (8.5). We can extend the sampling distribution $p(x|\xi^k) := \sum_{\alpha\in\Lambda^k} \varphi_\alpha^k \cdot \tilde{p}(x|\theta_k^\alpha) \in \mathcal{L}_{V^k}$ to a distribution $\hat{p}(x|\xi^k) \in \mathcal{L}_{\hat{V}^{k+1}}$ over the interim search region $\hat{V}^{k+1}$ by:

$$
\begin{aligned}
\hat{p}(x|\xi^k) &:= \sum_{\alpha\in\hat{\Lambda}^k} \varphi_\alpha^k \cdot \tilde{p}(x|\theta_k^\alpha) = \sum_{\alpha\in\Lambda^k} \varphi_\alpha^k \cdot \tilde{p}(x|\theta_k^\alpha) + \sum_{\alpha\in\hat{\Lambda}^k\setminus\Lambda^k} \varphi_\alpha^k \cdot \tilde{p}(x|\theta_k^\alpha) \\
&= \sum_{\alpha\in\Lambda^k} \varphi_\alpha^k \cdot \tilde{p}(x|\theta_k^\alpha) + 0 = p(x|\xi^k) \quad ,
\end{aligned}
$$

where the last line is due to $\varphi_\alpha^k = 0$ for $\alpha \in \hat{\Lambda}^k \setminus \Lambda^k$. The expected fitness of the $k$th iteration on the interim search region $\hat{V}^{k+1}$ is thus given by:

$$
\begin{aligned}
E^k &= \int_{x\in V^k} f(x)p(x|\xi^k)dx = \int_{x\in\hat{V}^{k+1}} f(x)\hat{p}(x|\xi^k)dx \\
&= \int_{\hat{V}^{k+1}} f(x) \sum_{\alpha\in\Lambda^k} \varphi_\alpha^k \cdot \tilde{p}(x|\theta_k^\alpha)dx + \int_{\hat{V}^{k+1}} f(x) \sum_{\alpha\in\hat{\Lambda}^{k+1}\setminus\Lambda^k} \varphi_\alpha^k \cdot \tilde{p}(x|\theta_k^\alpha)dx \\
&= \sum_{\alpha\in\Lambda^k} \varphi_\alpha^k \cdot E_\alpha^k + 0 \quad ,
\end{aligned}
\tag{8.12}
$$

where $E_\alpha^k := \int_{\exp_{x_\alpha}(B(\mathbf{0}, j_{x_\alpha}))} f(x)\tilde{p}(x|\theta^\alpha)dx$ for $\alpha \in \Lambda^k$ for simplicity (see Eq. (8.4)).

Similarly, let $X^{k+1} \subset X^k \cup \hat{X}^{k+1}$ be the initial search centroids of the $(k+1)$th iteration, indexed by $\Lambda^{k+1} \subset \hat{\Lambda}^{k+1}$, corresponding to search region $V^{k+1} \subset \hat{V}^{k+1}$. We can extend the sampling distribution of the $(k+1)$th iteration, denoted by $p(x|\xi^{k+1}) := \sum_{\alpha\in\Lambda^{k+1}} \varphi_\alpha^{k+1} \cdot \tilde{p}(x|\theta_{k+1}^\alpha) \in \mathcal{L}_{V^{k+1}}$ to a distribution $\hat{p}(x|\xi^{k+1}) \in \mathcal{L}_{\hat{V}^{k+1}}$ over the interim search region $\hat{V}^{k+1}$ by:

$$
\begin{aligned}
\hat{p}(x|\xi^{k+1}) &:= \sum_{\alpha\in\hat{\Lambda}^{k+1}} \varphi_\alpha^{k+1} \cdot \tilde{p}(x|\theta_{k+1}^\alpha) = \sum_{\alpha\in\Lambda^{k+1}} \varphi_\alpha^{k+1} \cdot \tilde{p}(x|\theta_{k+1}^\alpha) + \sum_{\alpha\in\hat{\Lambda}^{k+1}\setminus\Lambda^{k+1}} \varphi_\alpha^k \cdot \tilde{p}(x|\theta_k^\alpha) \\
&= \sum_{\alpha\in\Lambda^{k+1}} \varphi_\alpha^{k+1} \cdot \tilde{p}(x|\theta_{k+1}^\alpha) + 0 = p(x|\xi^{k+1}) \quad .
\end{aligned}
$$

Analogously, the expected fitness of $(k+1)$th iteration can be found in the interim search region $\hat{V}^{k+1}$:

$$
\begin{aligned}
E^{k+1} &= \int_{x \in V^{k+1}} f(x) p(x|\xi^{k+1}) dx = \int_{x \in \hat{V}^{k+1}} f(x) \hat{p}(x|\xi^{k+1}) dx \\
&= \int_{\hat{V}^{k+1}} f(x) \sum_{\alpha \in \Lambda^{k+1}} \varphi_\alpha^{k+1} \cdot \tilde{p}(x|\theta_{k+1}^\alpha) dx + \int_{\hat{V}^{k+1}} f(x) \sum_{\alpha \in \hat{\Lambda}^{k+1} \setminus \Lambda^{k+1}} \varphi_\alpha^{k+1} \cdot \tilde{p}(x|\theta_{k+1}^\alpha) dx \\
&= \sum_{\alpha \in \Lambda^{k+1}} \varphi_\alpha^{k+1} \cdot E_\alpha^{k+1} + 0 \quad ,
\end{aligned}
$$

By the selection step of line 6 of Algorithm 8.2, new centroids $X^{k+1} := \{x_\alpha\}_{\alpha \in \Lambda_{k+1}}$ are selected based on the corresponding expected fitness. This implies for any $\beta \in \Lambda^{k+1}, \alpha \in \Lambda^k$, we have: $E_\beta^{k+1} \geq E_\alpha^k$. By Eq. (8.12), since $\sum_{\alpha \in \Lambda^k} \varphi_\alpha = 1$, $\sum_{\beta \in \Lambda^{k+1}} \varphi_\beta = 1$ and $|\Lambda^{k+1}| = |\Lambda^k|$, we obtain the following inequality on the interim search region $\hat{V}^{k+1}$:

$$
E^{k+1} = \sum_{\beta \in \Lambda^{k+1}} \varphi_\beta \cdot E_\beta^{k+1} \geq \sum_{\alpha \in \Lambda^k} \varphi_\alpha \cdot E_\alpha^k = E^k \quad ,
$$

as desired. □

### 8.2.6 Exploration Distribution of Extended RSDFO

In this section we describe the exploration distribution, denoted by $U$, in line 2 of Algorithm 8.2. At the $k$th iteration, the exploration distribution $U$ is defined to be the uniform distribution on the boundary of the *explored region up-to the $k$th iteration*. This allows us to explore new search regions in the manifold search space $M$ beyond the current explored region using only local computations.

For each $k > 0$, recall from Eq. (8.5), the interim search region $\hat{V}^{k+1}$ denoting the closed search region explored in the $k$th iteration of Extended RSDFO. Consider for each $k$, the union of *all* search regions of $M$ explored by the algorithm *up-to* the $k$th iteration, given by:

$$
W^k = \bigcup_{j=0}^{k} \hat{V}^{j+1} \quad . \tag{8.13}
$$

The **exploration distribution** $U$ in line 2 of Algorithm 8.2 is given by the uniform distribution over the *boundary* of the explored region *up-to* the $k$th iteration:

$$
U := \text{unif}\left(\partial\left(W^k\right)\right) \quad .
$$

Points from the exploration distribution $U$ can be generated by an acceptance-rejection approach using local computations on the tangent spaces, this is illustrated in Fig. 8.4.

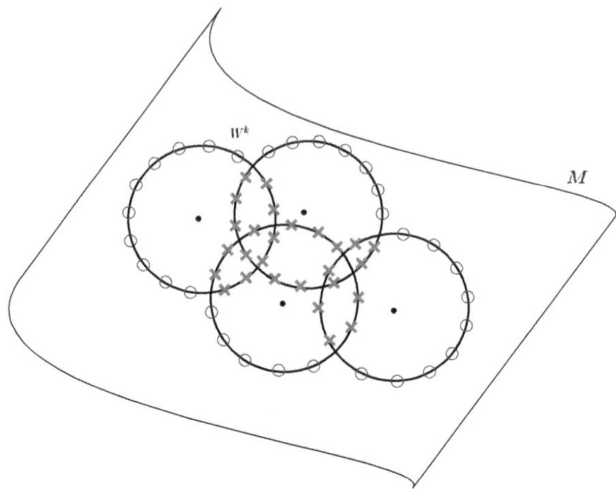


**Fig. 8.4** An simplified illustration of the exploration distribution sampling using acceptance-rejection method. We first sample uniformly from the boundary of each of the geodesic balls, with centroids labelled by the filled circles. Points sampled from the boundary of the geodesic balls that are also in the interior of $W^k$ will be close to *at least one* of the centroids, hence are discarded according to Eq. (8.16). Therefore points labelled with hollow circles are accepted, whereas the points labelled by crosses are rejected

More formally, using the notation from Sect. 8.2.4, summarized in Table 8.1: for $k > 0$, let $X^k \cup \hat{X}^{k+1}$ denote the set of centroids explored in the $k$th iteration of Extended RSDFO. The set of **accumulated centroids** explored by Extended RSDFO *up-to* the $k$th iteration (this includes the explored but discarded ones in the interim search centroids) is given by:

$$X_A^k := \bigcup_{j=0}^{k} X^j \cup \hat{X}^{j+1} \quad . \tag{8.14}$$

The set of accumulated centroids $X_A^k$ is indexed by the finite set $\Lambda_A^k := \cup_{j=0}^k \hat{\Lambda}^{j+1}$.

To generate points on the boundary of the explored region, we first sample points from the union of the boundary of explored geodesic balls (the geodesic spheres):

$$y \sim \text{unif}\left(\bigcup_{\alpha \in \Lambda_A^k} \partial\left(\tilde{B}_\alpha\right)\right) \quad , \tag{8.15}$$

where $\tilde{B}_\alpha = \exp_{x_\alpha}\left(B(\mathbf{0}, j_{x_\alpha})\right)$ for simplicity. This sampling can be done with strictly local computations on the geodesic sphere $\partial\left(\tilde{B}_\alpha\right)$ within the tangent space $T_{x_\alpha}M$ around each accumulated centroids $x_\alpha \in X_A^k$ on the manifold.

It remains to reject points that lies in the interior of $W^k$. Since every geodesic ball in $M$ is also a metric ball in $M$ under Riemannian distance with the same radius [Lee06]: for any $x_\alpha \in X_A^k$, a point $y \in \tilde{B}_\alpha$ satisfies:

$$d(x_\alpha, y) \leq j_{x_\alpha} \leq \text{inj}(x_\alpha) \quad ,$$

with equality if only if $y \in \partial\left(\tilde{B}_\alpha\right)$, where $d$ denote the Riemannian distance function[6] [Lee06] on $M$ . Therefore to preserve sampled points on the boundary of the explored region, we reject $y$ that lies in the interior of *any* geodesic ball. In particular from Eq. (8.15) we reject $y$ if it satisfies: there exists $x_\alpha \in X_A^k$ such that

$$d(x_\alpha, y) < j_{x_\alpha} \quad . \tag{8.16}$$

### 8.2.7 Termination Criterion

Due to the "multi-layer" structure of Extended RSDFO (imposed by the product manifold structure of the decision space, see Fig. 8.2), we may choose different termination criteria depending on the usage of the algorithm. In this section we discuss two examples of termination criteria.

For practical implementations (for example in the experiments described in Chap. 9), due to limitations of computation power: we say that Extended RSDFO terminates when the local RSDFO's terminate in their own accord. In other words, as local RSDFO's inherit the termination criteria from their pre-adapted counterparts, we say that Extended RSDFO terminates when *all* the local RSDFO steams terminate around the current search centroids.

For the theoretical studies (for geometry of evolutionary step in Sect. 8.3.2, and convergence in Sects. 8.4 and 8.4.1), when $M$ is a compact connected Riemannian manifold: Extended RSDFO terminates when the boundary of the explored region (see Sect. 8.2.6) is empty.

## 8.3 Geometry of Evolutionary Step of Extended RSDFO

In this section, we consider a special case of Extended RSDFO. The aim is to discuss and flesh out the geometry and dynamics of the evolutionary step in Extended RSDFO (Algorithm 8.2) discussed in this chapter.

Let $M$ be a Riemannian manifold, the special case of Extended RSDFO differs from the general case in one aspect: the local component densities around the search centroids in $M$ are fixed. In particular, the local RSDFO will determine new search

[6] This can be computed locally within the normal neighbourhood of $x_\alpha$.

centroids *without* changing the statistical parameter around the search centroids. This allows us to focus on the dynamics of mixture coefficients without considering the mixture component parameters generated/altered by the RSDFO core.

**Remark 8.1** We may assume the radius of geodesic ball $j_{x_\alpha} \ll \text{inj}(x_\alpha)$ is sufficiently small for all $\alpha \in \Lambda_M$, such that varying the mixture coefficients while keeping the mixture components density fixed will describe the overall density in sufficient detail. Suppose if we cover $M$ with sufficient number of mixture components, while fixing the individual mixture components and allowing the mixture coefficients to vary. The mixture density formed by fixed mixture components and free mixture coefficients will sufficiently describe the search distributions appropriate for our optimization problem. This is illustrated in Fig. 8.5.

More formally, using the notation from Sect. 8.2.1: Let $\mathcal{E}_M$ denote an orientation-preserving open cover of $M$ by geodesic balls indexed by $\Lambda_M$. Let $\Lambda_V \subset \Lambda_M$ be a finite subset of indices, and let $V$ denote a typical search region of Extended RSDFO define by the set of centroids $\{x_\alpha\}_{\alpha \in \Lambda_V}$ and corresponding geodesic balls as described in Eq. (8.1). Let $\mathcal{E}_V \subset \mathcal{E}_M$, denote the finite sub-cover of $\mathcal{E}_M$ over $V$.

We begin the discussion by constructing a special case of the aforementioned (see Sect. 8.2.1) mixture family of densities where families of local component densities consisting of a *single* fixed density: i.e. $S_\alpha = \{F_\alpha\}$ for each $\alpha \in \Lambda_V$. For

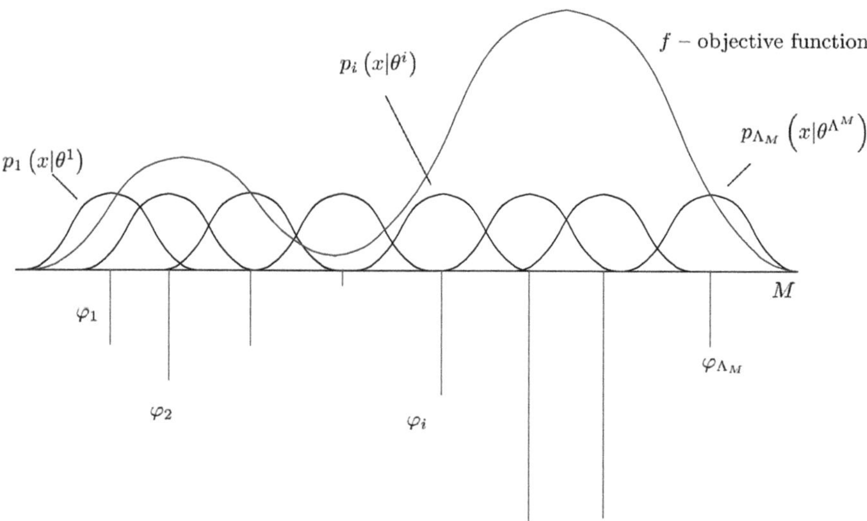


**Fig. 8.5** A simplified illustrated of the mixture distribution in the special case if the search region is sufficiently well covered. The figure shows a 1-dimensional manifold $M$, covered by fixed mixture components and free mixture coefficients $\varphi_\alpha$. The promising regions in $M$ corresponding to the fitness landscape of objective function $f$ can be well expressed through the search distributions obtained solely by varying mixtures coefficients. The magnitude of mixture coefficients $\varphi$'s are showed in the figure above as downward bars

example, for each $\alpha \in \Lambda_V$, we may let $F_\alpha := \overline{N}(0, I)$ denote the restriction of the spherical Gaussian density $N(0, I)$ to $B(0, j_{x_\alpha}) \subset T_{x_\alpha}M$ (properly renormalized). More generally, let the singleton $S_\alpha = \{F_\alpha\}$ denote the local component density over $B\left(\mathbf{0}, j_{x_\alpha}\right) \subset T_{x_\alpha}M$. For each $\alpha \in \Lambda_V$, the inherited component density over $\exp_{x_\alpha}\left(B\left(\mathbf{0}, j_{x_\alpha}\right)\right)$ consists of a single element: $\tilde{S}_\alpha = \left\{\tilde{F}_\alpha := \exp_{x_\alpha}^{-1^*} F_\alpha\right\}$.

The family of parametrized mixture densities on $M$ described in Eq. (8.3) thus reduces to the following special case:

$$\overline{\mathcal{L}}_V := \left\{ \sum_{\alpha \in \Lambda_V} \varphi_\alpha \cdot \tilde{F}_\alpha \,\middle|\, \{\varphi_\alpha\}_{\alpha \in \Lambda_V} \in \overline{S}_0,\ \tilde{F}_\alpha \in \tilde{S}_\alpha \right\} \quad , \tag{8.17}$$

where $\overline{S}_0$ denote the closure of simplex described in Eq. (8.2). The set of mixture densities $\overline{\mathcal{L}}_V$ over $V$ is therefore the product Riemannian manifold given by $\overline{\mathcal{L}}_V = \overline{S}_0 \times \bigoplus_{\alpha \in \Lambda_V} \tilde{S}_\alpha$, which in turn is diffeomorphic to $\overline{S}_0$. In this sense, this special case allows us to focus on the dynamics of mixture coefficients in $\overline{S}_0$.

In this section we propose a modification of the Fisher information metric to describe the geometry of $\overline{\mathcal{L}}_V \cong \overline{S}_0$. The proposed metric is used in **both** Extended RSDFO described in the previous section and the following special case. There are two reasons to use the modified metric on the closure of the simplex $\overline{S}_0$ as oppose to the Fisher information metric:

1. Fisher information metric is defined only on the interior of the simplex $S_0$, not on the closure of the simplex $\overline{S}_0$.
2. Natural gradient ascent/descent [CM87, Ama98] on the simplex $S_0$ under Fisher information metric favours the vertices of the simplex, driving the search towards local extremum. On the other hand, Natural gradient ascent on the closure $\overline{S}_0$ under the modified metric, as we will discuss in the subsequent discussion, favours the interior point of the simplex with coordinates proportional to the relative weights of the vertices.

We now make explicit the geometrical structure of the modified metric on $\overline{S}_0$ as follows:

**Definition 8.1** Let $F$ denote the Fisher information matrix on the mixture coefficient simplex $S_0$ [Leb02]. Let $\epsilon_0 > 0$ be an arbitrarily small non-zero real number, consider the matrix:

$$G := F^{-1} + \epsilon_0 \cdot I \quad . \tag{8.18}$$

Let $\xi := \{\varphi_\alpha\}_{\alpha \in \Lambda_V}$ be the local coordinate of a point in the *closure* of the mixture coefficient simplex $\overline{S}_0 \cong \overline{\mathcal{L}}_V$, and let $\{\partial^\alpha\}_{\alpha \in \Lambda_V}$ denote the corresponding local coordinate frame on $T_\xi \overline{S}_0$. Let $Y := \sum_{\alpha \in \Lambda_V} y_\alpha \partial^\alpha$, $Z := \sum_{\alpha \in \Lambda_V} z_\alpha \partial^\alpha$ be vector fields over $\overline{S}_0$. The **simplicial** $\epsilon_0-$**metric** $g_{\epsilon_0}$ is the Riemannian metric on $\overline{S}_0$ corresponding to the matrix $G$ given by:

$$g_{\epsilon_0}(Y, Z) := \sum_{\alpha \in \Lambda_V} y_\alpha z_\alpha \cdot (\varphi_\alpha + \epsilon_0) \quad . \tag{8.19}$$

The canonical divergence on $\overline{\mathcal{L}}_V \cong \overline{S}_0$ with respect to the simplicial $\epsilon_0$−metric $g_{\epsilon_0}$ is the Bregman divergence [AC10] given by the strictly convex function on the natural parameters of $\overline{S}_0$:

$$\begin{aligned} \kappa : \overline{\mathcal{L}}_V \cong \overline{S}_0 &\to \mathbb{R} \\ \xi = (\varphi_\alpha)_{\alpha \in \Lambda_V} &\mapsto \frac{1}{6} \sum_{\alpha \in \Lambda_V} (\varphi_\alpha + \epsilon_0)^3 \quad , \end{aligned}$$

hence the canonical divergence on $\overline{\mathcal{L}}_V$ is given by:

$$\begin{aligned} D : \overline{\mathcal{L}}_V \times \overline{\mathcal{L}}_V \cong \overline{S}_0 \times \overline{S}_0 &\to \mathbb{R} \\ (\xi, \xi') &\mapsto \kappa(\xi) - \kappa(\xi') - \langle \nabla\kappa(\xi'), \varphi - \xi' \rangle \quad . \end{aligned} \tag{8.20}$$

The dualistic structure of the statistical manifold $\overline{\mathcal{L}}_V \cong \overline{S}_0$, induced by the Bregman divergence described above, is in fact dually flat. Interested readers may find the general construction of the dualistic structure of $\overline{\mathcal{L}}_V \cong \overline{S}_0$ under the Bregman divergence described above in [AC10].

Furthermore, let the function $f : M \to \mathbb{R}$ denote the objective function of an optimization problem over Riemannian manifold $M$. Let $\xi := \{\varphi_\alpha\}_{\alpha \in \Lambda_V}$ denote the set of parameters of $\overline{\mathcal{L}}_V$, then for each $p(x|\xi) := \sum_{\alpha \in \Lambda_V} \varphi_\alpha \tilde{F}_\alpha \in \overline{\mathcal{L}}_V \subset \text{Prob}(V)$, the expected fitness [WSPS08b] of $f$ over $p(x|\xi) \in \overline{\mathcal{L}}_V$ (supported in set $V \subset M$) in the special case is given by:

$$\begin{aligned} J^V(\xi) &:= \int_{\text{supp}(p(\xi))} f(x) p(x|\xi) dx \\ &= \int_{\text{supp}(p(\xi))} f(x) \sum_{\alpha \in \Lambda_V} \varphi_\alpha \cdot \tilde{F}_\alpha(x) dx \quad . \end{aligned} \tag{8.21}$$

In practice, this integral can be approximated using $p(x|\xi)$−Monte Carlo samples for $x$ on the search space manifold $M$.

### 8.3.1 *Geometry and Simplicial Illustration of Evolutionary Step*

By the product Riemannian structure of family of mixture densities $\overline{\mathcal{L}}_V$, the change of mixture coefficients in the evolutionary step of Extended RSDFO (Algorithm 8.2) is *independent* from the change of the parameter of the mixture component densities.

Moreover, in the special case described above, the family of mixture component densities consists of a singleton: $\tilde{S}_\alpha = \left\{ \tilde{F}_\alpha := \exp_{x_\alpha}^{-1^*} F_\alpha \right\}$, and there will not be any changes in the parameter of the component densities. Therefore it remains to consider the changes in mixture coefficients in $\overline{S}_0$.

By the discussion of Sect. 8.2.4, for each iteration of Extended RSDFO, we transverse through three families of parametrized mixture densities on $M$:

$$\overline{\mathcal{L}}_V = \left\{ \sum_{\alpha \in \Lambda_V} \varphi_\alpha \cdot \tilde{F}_\alpha \,\middle|\, \{\varphi_\alpha\}_{\alpha \in \Lambda_V} \in \overline{S}_0,\ \tilde{F}_\alpha \in \tilde{S}_\alpha \right\} \quad , \tag{8.22}$$

where the search region $V$ and the corresponding centroids $X$, index sets $\Lambda_V$, and simplices $\overline{S}_0$ are summarized in Table 8.1 with $\overline{\mathcal{L}}_V$ replacing $\mathcal{L}_V$. Note that in this case for all $\alpha \in \Lambda_V$, where $\Lambda_V = \Lambda^k, \hat{\Lambda}^{k+1}, \Lambda^{k+1}$, the set of inherited density $\tilde{S}_\alpha = \left\{ \tilde{F}_\alpha := \exp_{x_\alpha}^{-1^*} F_\alpha \right\}$ consists of a single density on $\tilde{B}_\alpha \subset M$ for each $\alpha$ (as described in the beginning of Sect. 8.3).

Consider a maximization problem over search space manifold $M$. The evolution of the mixture coefficients in $\overline{\mathcal{L}}_V \cong \overline{S}_0$ is thus split into the following three parts, and the corresponding geometrical implications is illustrated in Fig. 8.6 (extending Fig. 8.3).

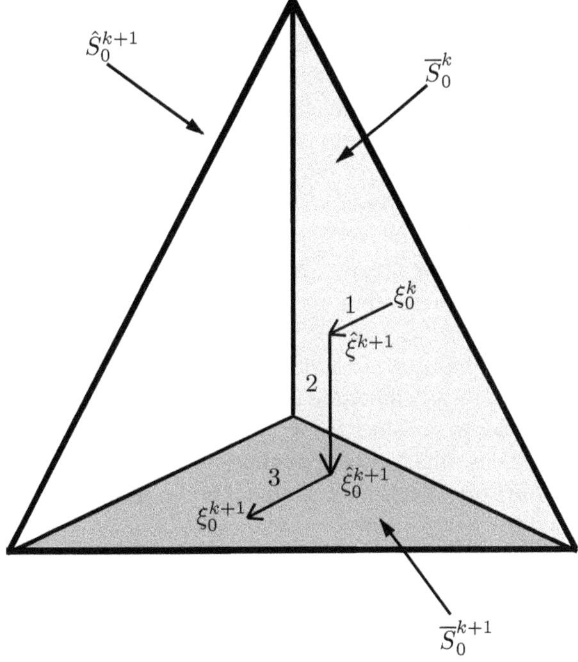


**Fig. 8.6** Simplicial illustration of the evolutionary step of Algorithm 8.2. The number on the arrows corresponds to the Part number of the evolutionary step. The interim simplex $\hat{S}_0^{k+1}$ is illustrated by the entire simplex with 4 vertices. The subsimplices $\overline{S}_0^k$ and $\overline{S}_0^{k+1}$ is given by the right and bottom subsimplices (shaded) with 3 vertices respectively

Part one: Determine the fixed point of natural gradient ascent[7] [CM87, Ama98] in $\hat{S}_0^{k+1}$ with respect to the Riemannian structure corresponding to the simplicial $\epsilon_0$−metric, described in Eq. (8.19), starting at $\xi_0^k \in \overline{S}_0^k \subset \hat{S}_0^{k+1}$. This is illustrated by arrow 1 in Fig. 8.6, and the fixed point is denoted by $\hat{\xi}^{k+1}$.

Part two: Project $\hat{\xi}^{k+1} \in \hat{S}_0^{k+1}$ onto $\overline{S}_0^{k+1}$, the $(N_{cull} - 1)$-dimensional faces of $\hat{S}_0^k$. This is illustrated by arrow 2 Fig. 8.6.

Part three: Determine the new fixed point $\xi_0^{k+1} \in \overline{S}_0^{k+1}$ of natural gradient ascent[8] [CM87, Ama98] in $\hat{S}_0^{k+1}$ with respect to the Riemannian structure corresponding to the simplicial $\epsilon_0$−metric on $\overline{S}_0^{k+1}$. This is labelled by arrow 3 on Fig. 8.6.

$\overline{S}_0^k$, $\overline{S}_0^{k+1}$ are two faces of $\hat{S}_0^{k+1}$ sharing *at most* $(N_{cull} + N_{random}) - N_{random} = N_{cull}$ vertices.[9] A summary of the notation can be found in Table 8.1.

### 8.3.2 Detailed Description of Evolutionary Step

In the previous section, we discussed geometrically the simplicial structure of the evolutionary step of the $k$th iteration of (the special case of) Extended RSDFO, which was summarized into three parts. In this section we fleshout the three part process of the evolutionary step in mathematical rigour.

Let $f : M \to \mathbb{R}$ denote the objective function of a *maximization* problem over Riemannian manifold $M$. We assume, without loss of generality, that $f$ is a strictly positive function by translation, i.e. $f(x) > 0$ for all $x \in M$. The case where we have a minimization problem is discussed in Remark 8.3.

By line 4 of Algorithm 8.2, at the $k$th iteration the algorithm produces a set of intermediate centroids $X^k \cup \hat{X}^{k+1} \subset M$. The search region of the $k$th iteration is thus given by the subset $\hat{V}^{k+1} := \bigcup_{\alpha \in \hat{\Lambda}^{k+1}} \exp_{x_\alpha}\left(B(0, j_{x_\alpha})\right)$ of $M$.

Let $\mathcal{E}_{\hat{V}^{k+1}} := \left\{\left(\exp_{x_\alpha}, B(0, j_{x_\alpha})\right)\right\}_{X_\alpha \in X^k \cup \hat{X}^{k+1}} \subset \mathcal{E}_M$ denote the finite subset of the orientation-preserving open cover $\mathcal{E}_M$ of $M$. Consider the "special case" family of mixture densities over $\hat{V}^{k+1}$ equipped with the corresponding simplicial $\epsilon_0$−metric, denoted by $\left(\overline{\mathcal{L}}_{\hat{V}^{k+1}}, \hat{g}_{\epsilon_0}^k\right)$, defined in Eqs. (8.17) and (8.19) respectively. In the special case, the locally inherited component densities consist of a single element, therefore it suffices to consider the dynamics on the simplex of mixture coefficients.

We describe the three parts of the evolutionary of mixture coefficient in detail:

**Part one**: determine the fixed point of the natural gradient ascent in $\hat{S}_0^{k+1}$ on the modified maximization problem [WSPS08a]:

---

[7] We perform natural gradient ascent for maximization problem and descent for minimization problem. For minimization case please refer to Remark 8.3.

[8] Similar to part one: we perform natural gradient ascent for maximization problem and descent for minimization problem. For minimization case, refer to Remark 8.3.

[9] This can be zero, as illustrated in Fig. 8.2.

$$\max_{x \in M} f(x) \mapsto \max_{\xi^k \in \hat{S}_0^k} J^{\hat{V}^{k+1}}\left(\xi^k\right) \quad ,$$

where $\xi^k := (\varphi_\alpha)_{\alpha \in \hat{\Lambda}^{k+1}}$ denote the set of parameters of $\overline{\mathcal{L}}_{\hat{V}^{k+1}} \cong \hat{S}_0^{k+1}$. $J^{\hat{V}^{k+1}}\left(\xi^k\right)$ is the expected fitness of $f$ over $p(x|\xi^k) \in \overline{\mathcal{L}}_{\hat{V}^{k+1}}$ described in Eq. (8.21).

Let $\xi_0^k := \begin{cases} \varphi_\alpha^k & \text{if } \alpha \in \Lambda^k \subset \hat{\Lambda}^{k+1} \\ 0 & \text{otherwise.} \end{cases} \in \overline{S}_0^k \subset \hat{S}_0^{k+1}$ denote the initial point on the subsimplex $\overline{S}_0^k$ of $\hat{S}_0^{k+1}$. Let $\tilde{\nabla}_{\xi^k} J^{\hat{V}^{k+1}}\left(\xi^k\right)$ denote the natural gradient of $J^{\hat{V}^{k+1}}\left(\xi^k\right)$ on $\overline{\mathcal{L}}_{\hat{V}^{k+1}}$, and $\text{proj}_{\hat{S}_0^k}$ the projection mapping onto the interim simplex of mixture coefficients $\hat{S}_0^k$ described above. We perform constraint natural gradient ascent [CM87, Ama98] on $J^{\hat{V}^{k+1}}\left(\xi^k\right)$ at $\xi_0^k$, summarized by Algorithm 8.3 below:

**Algorithm 8.3:** Natural Gradient Ascent

**Data**: Initial point $\xi_0^k \in \overline{S}_0^k \subset \hat{S}_0^{k+1}$, objective function $J^{\hat{V}^{k+1}}\left(\xi^k\right)$, step size $s_i$. Set iteration count $i = 0$

1 initialization;
2 **while** *stopping criterion not satisfied* **do**
3 $\quad \overline{\xi}_{i+1}^k = \xi_i^k + s_i \cdot \tilde{\nabla}_{\xi^k} J^{\hat{V}^{k+1}}\left(\xi_i^k\right)$ ;
4 $\quad \xi_{i+1}^k = \text{proj}_{\hat{S}_0^{k+1}}\left(\overline{\xi}_{i+1}^k\right)$ ;
5 $\quad$ i = i+1;

Given sufficiently small step sizes $s_i > 0$, the fixed point of Algorithm 8.3 can be determined explicitly.[10] In particular, a point $\xi_i^k \in \hat{S}_0^{k+1}$ is the fixed point of (lines 3 and 4 of) Algorithm 8.3 if and only if $\text{proj}_{\hat{S}_0^{k+1}}\left(\xi_i^k + s_i \cdot \tilde{\nabla}_{\xi^k} J^{\hat{V}^{k+1}}\left(\xi^k\right)\right) = \xi_i^k$.

The above condition is satisfied if and only if $\tilde{\nabla}_{\xi^k} J^{\hat{V}^{k+1}}\left(\xi_i^k\right)$ is parallel to the normal vector: $[1, 1, \ldots, 1]$ of $\hat{S}_0^{k+1}$ of the interim simplex $\hat{S}_0^{k+1}$.

Therefore to determine the fix point, we perform the following computation: for $k > 0$, at the $k$th iteration of the special case of Extended RSDFO, the inverse of interim Riemannian metric matrix (Eq. (8.18)) $\hat{G}_k^{-1}$ corresponding to the metric $\hat{g}_{\epsilon_0}^k$ at a point $\xi_i^k = \left\{\varphi_\alpha^k\right\}_{\alpha \in \hat{\Lambda}^{k+1}}$ on the interim simplex $\hat{S}_0^{k+1} \cong \overline{\mathcal{L}}_{\hat{V}^{k+1}}$ is given by:

[10] The step sizes $s_i$'s are chosen to be sufficiently small such that the orthogonal projection $\text{proj}_{\hat{S}_0^{k+1}}\left(\overline{\xi}_{i+1}^k\right)$ along the normal vector $[1, 1, \ldots, 1] = \mathbf{1}$ of $\hat{S}_0^{k+1}$ onto the simplex $\hat{S}_0^{k+1}$ remains in the $\hat{S}_0^{k+1}$.

$$\hat{G}_k^{-1} := \begin{bmatrix} \frac{1}{\varphi_1^k+\epsilon_0} & 0 & \cdots & \cdots & 0 \\ 0 & \frac{1}{\varphi_2^k+\epsilon_0} & 0 & \cdots & 0 \\ \vdots & \ddots & \ddots & \ddots & 0 \\ 0 & \cdots & 0 & \frac{1}{\varphi_{\hat{\Lambda}^{k+1}-1}^k+\epsilon_0} & 0 \\ 0 & 0 & \cdots & 0 & \frac{1}{\varphi_{\hat{\Lambda}^{k+1}}^k+\epsilon_0} \end{bmatrix} . \tag{8.23}$$

The natural gradient [Ama98] $\tilde{\nabla}_{\xi^k} J^{\hat{V}^{k+1}}\left(\xi_i^k\right)$ is therefore given by:

$$\begin{aligned}
\tilde{\nabla}_{\xi^k} J^{\hat{V}^{k+1}}\left(\xi_i^k\right) &= \hat{G}_k^{-1} \cdot \nabla_{\xi^k} J^{\hat{V}^{k+1}}\left(\xi_i^k\right) \\
&= \hat{G}_k^{-1} \cdot \left(\frac{\partial}{\partial \varphi_\alpha^k} \int_x f(x) \sum_{\alpha \in \hat{\Lambda}^{k+1}} \varphi_\alpha^k \cdot \tilde{F}_\alpha(x) dx\right)_{\alpha \in \hat{\Lambda}^{k+1}} \\
&= \hat{G}_k^{-1} \cdot \left(\int_x f(x) \tilde{F}_\alpha(x) dx\right)_{\alpha \in \hat{\Lambda}^{k+1}} \\
&= \hat{G}_k^{-1} \cdot \left(E_\alpha^k [f]\right)_{\alpha \in \hat{\Lambda}^{k+1}} \\
&= \left(\frac{1}{\varphi_\alpha^k + \epsilon_0} \cdot E_\alpha^k [f]\right)_{\alpha \in \hat{\Lambda}^{k+1}} .
\end{aligned}$$

For simplicity and when the context is clear, for $\alpha \in \hat{\Lambda}^{k+1}$ we let $E_\alpha := E_\alpha^k [f]$ denote the expected fitness of $f$ over the geodesic ball $\exp_{x_\alpha}\left(B(0, j_{x_\alpha})\right)$ on the $k$th iteration of Extended RSDFO (see Eq. (8.4)). Since $f$ is strictly positive, we may assume by translation that $E_\alpha > 0$ for all $\alpha$. Therefore the natural gradient $\tilde{\nabla}_{\xi^k} J^{\hat{V}^{k+1}}\left(\xi_i^k\right)$ is parallel to **1** if and only if for all $\alpha \in \hat{\Lambda}^{k+1}$:

$$\frac{1}{\varphi_\alpha^k + \epsilon_0} \cdot E_\alpha = c \cdot 1 , \; c \in \mathbb{R}_+ \; . \tag{8.24}$$

Assume without loss of generality that $c = 1$, the above condition is then satisfied if and only if $\varphi_\alpha^k = E_\alpha - \epsilon_0$, for $\alpha \in \hat{\Lambda}^{k+1}$. Moreover, since the mixture coefficients $\left(\varphi_\alpha^k\right)_{\alpha \in \hat{\Lambda}^{k+1}} \in \hat{S}_0^{k+1}$ belong to a simplex, they must sum to 1: $\sum_{\alpha \in \hat{\Lambda}^{k+1}} \varphi_\alpha^k = 1$. The renormalized set of mixture coefficients thus becomes:

$$\begin{aligned}
\varphi_\alpha^k &= \frac{E_\alpha - \epsilon_0}{\sum_{\alpha \in \hat{\Lambda}^{k+1}} (E_\alpha - \epsilon_0)} \\
&= \frac{E_\alpha - \epsilon_0}{\sum_{\alpha \in \hat{\Lambda}^{k+1}} (E_\alpha) - |\hat{\Lambda}^{k+1}| \cdot \epsilon_0} .
\end{aligned} \tag{8.25}$$

A demonstration of the natural gradient ascent of **Part One** (and **Part three**) on the interim simplex $\hat{S}_0^{k+1}$ (and $\overline{S}_0^{k+1}$) is give in Fig. 8.7:

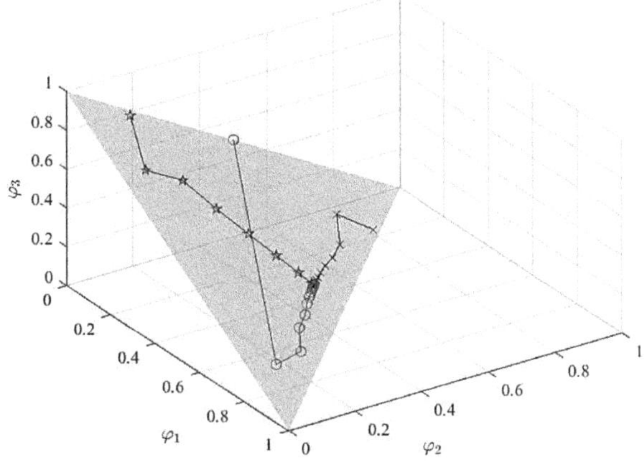


**Fig. 8.7** A demonstration of natural gradient ascent under the metric $G$ (Eq. (8.18)) in Part One of evolutionary step of Algorithm 8.2. The shaded surface represents the interim simplex $\hat{S}_0^{k+1}$ with three vertices $\varphi_1, \varphi_2, \varphi_3$. This search paths in the figure illustrates how the fixed point $\hat{\xi}^{k+1}$ of natural gradient ascent under $G$ is attained regardless of where the initial point is located in the simplex, whether it is an interior point or a boundary point

**Part two**: Let $\hat{\xi}^{k+1} := \left(\varphi_\alpha^k\right)_{\alpha\in\hat{\Lambda}^{k+1}} = \left(\frac{E_\alpha-\epsilon_0}{\sum_{\alpha\in\hat{\Lambda}^{k+1}}(E_\alpha-\epsilon_0)}\right)_{\alpha\in\hat{\Lambda}^{k+1}}$ denote the renormalized interim point described in Eq. (8.25) above. We now project this point to a $(N_{cull}-1)$-dimensional subsimplex $\overline{S}_0^{k+1}$ of $\hat{S}_0^{k+1}$.

We first sort the interim mixture coefficients $\hat{\xi}^{k+1} := \left(\varphi_\alpha^k\right)_{\alpha\in\hat{\Lambda}^{k+1}}$ in ascending order (or descending order for minimization problem), and let the indices $\Lambda^{k+1} \subset \hat{\Lambda}^{k+1}$ denote the largest (or smallest for minimization problem) $N_{cull}$ elements. The renormalized interim point is thus projected to the subsimplex $\overline{S}_0^{k+1}$ defined by:

$$\xi_0^k := \begin{cases} \frac{\varphi_\alpha^k}{\sum_{\alpha\in\Lambda^{k+1}}\varphi_\alpha^k} & \text{if } \alpha \in \Lambda^{k+1} \subset \hat{\Lambda}^{k+1} \\ 0 & \text{otherwise.} \end{cases} \in \overline{S}_0^{k+1} \subset \hat{S}_0^{k+1} \quad .$$

The subsimplex $\overline{S}_0^{k+1}$ therefore represents the set of mixture coefficients of the next iteration.

**Part three**: Finally, we determine the fixed point of natural gradient ascent (Algorithm 8.3) on the subsimplex $\overline{S}_0^{k+1} \subset \hat{S}_0^{k+1}$ with the projected interim point $\xi_0^k$ as the initial point.

By the derivations in part one, we compute the fixed point of the natural gradient ascent in the subsimplex $\overline{S}_0^{k+1}$ in a similar fashion. The inverse of Riemannian metric matrix $G_{k+1}^{-1}$ corresponding to the metric $g_{\epsilon_0}^{k+1}$ at a point $\xi_i^{k+1} = \left\{\varphi_\alpha^k\right\}_{\alpha\in\Lambda^{k+1}}$ on *the subsimplex* $\overline{S}_0^{k+1} \subset \hat{S}_0^{k+1}$ is given by:

$$G_{k+1}^{-1} := \begin{bmatrix} \frac{1}{\varphi_1^k+\epsilon_0} & 0 & \cdots & \cdots & 0 \\ 0 & \frac{1}{\varphi_2^k+\epsilon_0} & 0 & \cdots & 0 \\ \vdots & \ddots & \ddots & \ddots & 0 \\ 0 & \cdots & 0 & \frac{1}{\varphi_{\Lambda^{k+1}-1}^k+\epsilon_0} & 0 \\ 0 & 0 & \cdots & 0 & \frac{1}{\varphi_{\Lambda^{k+1}}^k+\epsilon_0} \end{bmatrix} ,$$

and the fixed point $\xi^{k+1} := \left(\varphi_\alpha^{k+1}\right)_{\alpha\in\Lambda^{k+1}}$ in $\overline{S}_0^{k+1}$ is therefore given by (similar to Eq. (8.25)):

$$\varphi_\alpha^{k+1} = \frac{E_\alpha - \epsilon_0}{\sum_{\alpha\in\Lambda^{k+1}} (E_\alpha - \epsilon_0)} \quad . \tag{8.26}$$

**Remark 8.2** The renormalized fixed point $\xi^{k+1}$ from Eq. (8.26) is an interior point of the simplex $\overline{S}_0^{k+1}$. Notice if the constant $0 < \epsilon_0 \ll 1$ is sufficiently small, such that it satisfies: $|\Lambda^{k+1}| \cdot \epsilon_0 \ll E_\alpha$ for all $\alpha \in \Lambda^{k+1}$, then we would retrieve Eq. (8.10):

$$\varphi_\alpha^{k+1} = \frac{E_\alpha - \epsilon_0}{\sum_{\alpha\in\Lambda^{k+1}} (E_\alpha - \epsilon_0)} \xrightarrow{|\Lambda^{k+1}|\cdot\epsilon_0 \ll E_\alpha} \frac{E_\alpha}{\sum_{\alpha\in\Lambda^{k+1}} E_\alpha} \quad .$$

The local coordinates of the fixed point described in Eq. (8.25) therefore directly corresponds to the relative expected fitness of each component $\exp_{x_\alpha}\left(B(0, j_{x_\alpha})\right)$ of the search region $V^{k+1} = \cup_{\alpha\in\Lambda^{k+1}} \exp_{x_\alpha}\left(B(0, j_{x_\alpha})\right)$.

Therefore the simplicial $\epsilon_0-$metric introduced in Eq. (8.19) describes the geometry of the evolutionary step of Extended RSDFO. In particular, the use of relative expected fitness in the evolution of mixture components of Eq. (8.10) is rigorously derived from first principles.

**Remark 8.3** In the case where we have a *minimization* problem over $M$, we would mirror the assumptions and operations of the maximization case. We assume, by translation, that the objective function is strictly negative: i.e. $f(x) < 0$ for all $x \in M$, which equivalently means $-f(x) > 0$ for all $x \in M$.

The natural gradient ascent in **Part One** (and similarly **Part three**) are replaced by natural gradient *descent* over the following converted minimization problem:

$$\min_{x\in M} f(x) \mapsto \min_{\xi^k\in\hat{S}_0^k} J^{\hat{V}^{k+1}}\left(\xi^k\right) \quad .$$

Since mixture coefficients $\left(\xi^k\right)$ in the simplices points towards the negative quadrant (see for example Fig. 8.7) under natural gradient *descent* direction, the normal vector of the simplices is taken to be $-\mathbf{1}$ instead. Equation (8.24) thus becomes:

$$\frac{1}{\varphi_\alpha^k + \epsilon_0} \cdot E_\alpha = c \cdot -1 \,, \; c \in \mathbb{R}_+ \quad .$$

The assumption that $f$ is strictly negative implies the expected value of $f$ over any geodesic ball $\exp_{x_\alpha}\left(B(0, j_{x_\alpha})\right)$ is strictly negative. In other words, the negative expected value is always positive: $-E_\alpha > 0$ for all $\alpha$, which in turn implies $\varphi_\alpha^k = -E_\alpha - \epsilon_0 > 0$. Equation (8.25) (similarly Eqs. (8.26) and (8.10)) will then become:

$$\varphi_\alpha^k = \frac{-E_\alpha - \epsilon_0}{\sum_{\alpha' \in \Lambda} (-E_\alpha - \epsilon_0)} \cong \frac{-E_\alpha}{\sum_{\alpha \in \Lambda} -E_\alpha} \quad . \tag{8.27}$$

Note that since we assumed $f(x)$ is strictly negative, $\varphi_\alpha^k$ can be equivalently expressed as $|E_\alpha|/(\sum_{\alpha \in \Lambda} |E_\alpha|)$. The rest of the arguments and computation are the same.

## 8.4 Convergence of Extended RSDFO on Compact Connected Riemannian Manifolds

In this section we discuss the convergence behavior of Extended RSDFO on compact connected Riemannian manifolds. In particular, we show how Extended RSDFO converges globally eventually within *finitely many* steps on compact connected Riemannian manifolds under the assumption that the optima is attainable in the manifold. We begin the discussion by an overview of the result; the formal definitions and detailed proof will be provided in Sect. 8.4.1.

Global convergence of SDFOs on Euclidean spaces, such as Estimation of Distribution Algorithms (EDA) on $\mathbb{R}^n$ studied in [ZM04], is derived based on the explicit relations between the sampling probability distribution on the $k$th and the $(k+1)$th iteration of the EDA algorithm. This relation in turn depends on the chosen selection scheme, and more importantly: the fact that the selection distributions share the same support in the Euclidean space. Similarly, convergence results of Evolutionary Strategies [Bey14] on Euclidean spaces share the same underlying assumption: selection distributions across iterations share the same support. As such, the selection distributions belong to the same statistical manifold.

However, such global convergence results of EDAs cannot be translated directly to Riemannian manifolds. Let $M$ be a Riemannian manifold and two points $x_k, x_{k+1}$ in $M$. The tangent spaces $T_{x_k}M$ and $T_{x_{k+1}}M$ centered at $x_k, x_{k+1}$ respectively are different (disjoint) spaces. Locally inherited parametrized probability densities (described in Chap. 5) over $T_{x_k}M$ and $T_{x_{k+1}}M$ have different supports and thus belong to different families. On the other hand, in Extended RSDFO, family of parametrized densities on separate tangent spaces can be related by mixture densities (described in Sect. 8.2.1), and in Proposition 8.1 the quality of solutions is shown to be monotonically non-decreasing from iteration to iteration.

Suppose a new boundary exploration point is generated for each iteration of Extended RSDFO, and suppose the algorithm terminates *only if* no boundary point is available (i.e. when the boundary of the explored region is empty). The explored region of Extended RSDFO consists of the union of geodesic balls in $M$ with non-zero radius. Let $j_M$ denote the smallest of the radii, then the sequence of "exploration centroids", generated on the boundary of the previous explored region, must be at least $j_M$ apart.

The sequence of boundary exploration points cannot continue indefinitely, otherwise this sequence will have no limiting point in $M$, contradicting the fact the $M$ is compact. Therefore the explored region generated by Extended RSDFO must exhaust the manifold $M$ in finitely many steps. Together with the use of elitist selection of the mixture components, the global optima will eventually be attainable in the explored region within finitely many steps.

### 8.4.1 Detailed Exposition of Convergence Behaviour of Extended RSDFO

In this section, we provide the detailed proof of the convergence of Extended RSDFO in connected compact Riemannian manifolds summarized the discussion in Sect. 8.4 above. In particular, we show that if Extended RSDFO generates a boundary exploration point for each iteration (Sect. 8.2.7), then it converges eventually globally in finitely many steps in compact connected Riemannian manifolds.

Let $M$ be a compact connected Riemannian manifold. For $k > 0$, let $X_A^k$ denote the finite set of accumulated centroids described in Eq. (8.14). Let $\mathcal{E}_M$ denote the orientation-preserving open cover of $M$ by geodesic balls described in Sect. 8.2.1 with set of centroids given by $X_M := \{x_\alpha\}_{\alpha\in\Lambda_M} \subset M$. The two sets of search centroids $X_M$ and $X_A^k$ describes an open cover of $M$:

**Definition 8.2** For each $k > 0$, let $X_M^k := X_M \cup X_A^k$. The **accumulative open cover** of $M$ described by $X_M^k$ is given by:

$$\mathcal{E}_M^k := \left\{\left(\exp_{x_\alpha}, B(\mathbf{0}, j_{x_\alpha})\right)\right\}_{\alpha\in\Lambda_M^k} ,$$

where $j_{x_\alpha} \leq \operatorname{inj}(x_\alpha)$.

Let $V^k$ denote the search region of the $k$th iteration of Extended RSDFO defined in Eq. (8.5). Using the notation of Table 8.1 in Sect. 8.2.4, $V^k$ is defined by the set of centroids $X^k \subset X_M^k$ indexed by the finite subset $\Lambda^k \subset \Lambda_M^k$.

Let the function $f : M \to \mathbb{R}$ denote the objective function of an optimization problem over $M$. Assume without loss of generality that the optimization problem is a maximization problem and $f$ is strictly positive. Furthermore, we assume that $f$ is bounded above in $M$ and that there exists an attainable global optima $x^* \in M$ such that: $f(x^*) \geq f(x)$ for all $x \in M$.

The existence of a global optima implies for all $k > 0$, there exists $\alpha^* \in \Lambda_M^k$ such that $E_{\alpha^*} := \int_{\exp_{x_{\alpha}^*}(B(\mathbf{0}, j_{x_{\alpha}^*}))} f(x)p(x|\theta^{\alpha^*})dx \geq E_\beta$ for all $\beta \in \Lambda_M$. We denote this by $E^* := E_{\alpha^*}$ for simplicity.

Motivated by the convergence condition of Euclidean SDFO described in the literature [ZM04], we outline a global convergence condition of Extended RSDFO:

**Definition 8.3** Extended RSDFO **converges globally eventually** if it satisfies the following condition:

$$\lim_{k \to \infty} \sup_{\alpha \in \Lambda^k} E_\alpha^k = E^* \quad , \tag{8.28}$$

where $E_\alpha^k := \int_{\exp_{x_\alpha}(B(\mathbf{0}, j_{x_\alpha}))} f(x)p(x|\theta^\alpha)dx$ for $\alpha \in \Lambda^k$.

This means asymptotically, the best component of the accumulative open cover $\mathcal{E}_M^k$ will be contained in the preserved centroids contained in $V^k$ with centroids $X^k$ indexed by $\Lambda^k$. In other words, there exists an integer $K_0$ such that for $k \geq K_0$ the set of retained centroids will contain the neighbourhood $\exp_{x_\alpha^*}\left(B(\mathbf{0}, j_{x_\alpha^*})\right)$ with the highest expected fitness.

We now show that, if Extended RSDFO terminates only when the boundary of the explored region is empty (Sect. 8.2.7), then Extended RSDFO convergence globally eventually (Definition 8.3) within finitely many steps on optimization problems over compact connected Riemannian manifolds.

**Lemma 8.1** *Let $(M, g)$ be a compact connected Riemannian manifold. Suppose for each iteration Extended RSDFO generates at least one exploration point from the boundary of the explored region*[11] *when the boundary is nonempty. Then Extended RSDFO explores $M$ in finitely many steps.*

***Proof*** For each $k \geq 0$, consider the region in $M$ explored by Extended RSDFO up to the $k$th iteration (see Eq. (8.13)):

$$W^k = \cup_{j=1}^k \hat{V}^j \subset M \quad .$$

$W^K$ is the finite union of closed geodesic balls in $M$, hence a nonempty closed subset of $M$. If $W^k \subsetneq M$, then the boundary $\partial\left(W^k\right)$ of $W^k$ must be nonempty: Since otherwise $W^k = \text{int}(W^k)$, meaning $W^k$ is a clopen proper subset of $M$, which contradicts the fact that $M$ is connected. This means we can always find a search centroid on the boundary $\partial\left(W^k\right)$ *unless* $W^k = M$.

Let $k \geq 0$, we generate a sequence of points $\left\{y^k\right\}_{k \in \mathbb{N}}$ in $M$ as follows:

(1) If $W^k \subsetneq M$, then by the argument above the boundary of the explored region is nonempty: $\partial\left(W^k\right) \neq \emptyset$. Hence by the assumption, Extended RSDFO can generate a new exploration centroid from $\partial\left(W^k\right)$. In particular, at the $k$th iteration, there exists $y^{k+1} \in X^k \cup \hat{X}^{k+1} \subset M$ in the set of interim centroids such that $y^{k+1}$

[11] See Sect. 8.2.6.

is generated from the boundary of explored region: i.e. $y^{k+1} \in \partial\left(W^k\right) \subset M$. Let $j_M := \inf_{x \in M} j_x \leq \inf_{x \in M} \text{inj}\,(x)$, then $j_M > 0$ and we obtain the following relation:

$$W^k \subset \exp_{y^{k+1}}\left(B(\mathbf{0}, j_M)\right) \cup W^k \subseteq W^{k+1} \quad ,$$

where $\exp_{y^{k+1}}\left(B(\mathbf{0}, j_M)\right)$ denotes the closed geodesic ball around $y^{k+1}$ with radius $j_M \leq j_{y^{k+1}} \leq \text{inj}\left(j_{y^{k+1}}\right)$. Note that geodesic balls in $M$ are also metric balls of the same radius [Lee06].
(2) Otherwise if $W^k = M$, then $\partial\left(W^k\right) = \emptyset$. We therefore set $y^{k+1} = y^k$ and $W^{k+1} = W^k = M$.

Since $M$ is compact, it admits the Bolzano-Weierstrass Property [R+64]: every sequence in $M$ has a convergent subsequence. In particular the sequence $\left\{y^k\right\}_{k \in \mathbb{N}}$ has a convergent subsequence $\left\{y^{k^\ell}\right\}_{k^\ell \in \mathbb{N}}$. This means there exists $\hat{y} \in M$ such that:

$$\hat{y} = \lim_{\ell \to \infty} y^{k^\ell} \quad .$$

Let $d_g : M \times M \to \mathbb{R}$ denote the Riemannian distance induced by the Riemannian metric $g$ of $M$. The above equation implies there exists finite $k^\alpha, k^\beta > 0$, $k^\alpha < k^\beta$ such that $d_g\left(y^{k^\alpha}, \hat{y}\right), d_g\left(y^{k^\beta}, \hat{y}\right) < \frac{j_M}{2}$, hence $d_g\left(y^{k^\alpha}, y^{k^\beta}\right) < j_M$.[12] Therefore we have:

$$y^{k^\beta} \in \text{int}\left(\exp_{y^{k^\alpha}}\left(B(\mathbf{0}, j_M)\right)\right) \subset \exp_{y^{k^\alpha}}\left(B(\mathbf{0}, j_M)\right) \subset W^{k^\alpha} \subset W^{k^\beta - 1} \quad .$$

which implies $y^{k^\beta} \in \text{int}\left(W^{k^\beta - 1}\right)$.

Finally, the result above means $W^{k^\beta - 1} = M$: Suppose by contrary that $W^{k^\beta - 1} \neq M$, then by construction of the sequence $\left\{y^k\right\}_{k \in \mathbb{N}}$ satisfies:

$$y^{k^\beta} \in \partial\left(W^{k^\beta - 1}\right) := W^{k^\beta - 1} \setminus \text{int}\left(W^{k^\beta - 1}\right) \quad ,$$

which contradicts the result above. Therefore there exists a finite integer $k^\beta > 0$ such that $W^{k^\beta - 1} = M$, as desired. □

**Theorem 8.1** *Let $(M, g)$ be a compact connected Riemannian manifold. Suppose for each iteration Extended RSDFO generates at least one exploration point from the boundary of the explored region[13] when the boundary is nonempty. Then Extended RSDFO satisfies the global convergence condition Eq. (8.28) on $M$ within finitely many steps. That is, there exists finite integer $N > 0$ such that:*

---

[12] In the normal neighbourhood, the Euclidean ball in the tangent space is mapped to the geodesic ball of the same radius.

[13] See Sect. 8.2.6.

$$\sup_{\alpha\in\Lambda^N} E_\alpha^N = \lim_{k\to\infty} \sup_{\alpha\in\Lambda^k} E_\alpha^k = E^* \quad ,$$

*where* $E_\alpha^k := \int_{\exp_{x_\alpha}(B(\mathbf{0}, j_{x_\alpha}))} f(x)p(x|\theta^\alpha)dx$ *for* $\alpha \in \Lambda^k$.

***Proof*** For each $k \geq 0$, consider the region in $M$ explored by Extended RSDFO up to the $k$th iteration (see Eq. (8.13)):

$$W^k = \cup_{j=1}^k \hat{V}^j \subset M \quad .$$

Let $\Lambda_A^k = \cup_{j=0}^k \hat{\Lambda}^{j+1}$ denote the index of accumulated centroids (see Eq. (8.14)). By Lemma 8.1, there exists finite integer $N > 0$ such that $W^N = M$. In particular, there exists $N > 0$ such that:

$$W^N = \lim_{k\to\infty} W^k = \lim_{k\to\infty} \bigcup_{\alpha\in\Lambda_A^k} \exp_{x_\alpha}\left(B(\mathbf{0}, j_{x_\alpha})\right) = M$$

Suppose $W^N$ is indexed by $\Lambda^N$. Since the fittest neighbourhoods are preserved by the evolutionary step (see Sect. 8.3.2, and Proposition 8.1) of Extended RSDFO, we obtain the following relation:

$$\begin{aligned}
\sup_{\alpha\in\Lambda^N} E_\alpha^N = \lim_{k\to\infty} \sup_{\alpha\in\Lambda^k} E_\alpha^k &= \lim_{k\to\infty} \sup_{\alpha\in\Lambda^k} \int_{\tilde{B}_\alpha\subset V^k} f(x)p(x|\theta^\alpha)dx \\
&= \lim_{k\to\infty} \sup_{\alpha\in\Lambda_A^k} \int_{\tilde{B}_\alpha\subset W^k} f(x)p(x|\theta^\alpha)dx \\
&= \sup_{\alpha\in\Lambda_A^k} \int_{\tilde{B}_\alpha\subset M} f(x)p(x|\theta^\alpha)dx \\
&= E^* \quad .
\end{aligned}$$

where $\tilde{B}_\alpha := \exp_{x_\alpha}\left(B(\mathbf{0}, j_{x_\alpha})\right)$ for simplicity. □

## 8.5 Discussion

Using the Riemannian adaption principle described in the previous chapter and the notion of locally inherited probability densities described in Chap. 5, we began the chapter by describing Riemannian Stochastic Derivative-Free Optimization (RSDFO), a generalized framework to translate SDFO algorithms from Euclidean spaces to Riemannian manifolds. Whilst RSDFO preserves the algorithmic structure and the local computations of the pre-adapted SDFO algorithms, it shares the notable shortcomings of manifold optimization algorithms in the literature: the computations and estimations are confined locally by a single normal neighbourhood, and the disjoint tangent spaces hinders further theoretical study of the properties of the algorithm.

To overcome the local restrictions of RSDFO, we required a parametrized probability densities defined beyond the normal neighbourhoods of Riemannian manifolds, and the mixture densities described in Chap. 6 provides exactly what is needed. The product Riemannian structure of parametrized mixture densities on Riemannian manifolds thus gave rise to a population-based stochastic meta-algorithm—Extended RSDFO based on the foundation of RSDFO. The product statistical Riemannian geometry of mixture densities over the search space manifold $M$ permeates all aspects of Extended RSDFO, which enables theoretical studies and geometrical insights into the intricacies of Extended RSDFO. In particular, we discussed the geometry and dynamics of the evolutionary steps of Extended RSDFO using a "regularized inverse" Fisher metric on the simplex of mixture coefficients, and showed that Extended RSDFO converges globally eventually in finitely many steps on connected compact Riemannian manifolds.

In the next chapter, we will wrap up the book by comparing Extended RSDFO against several state-of-the-art manifold optimization methods in the literature, such as Riemannian Trust-Region method [ABG07, AMS09], Riemannian CMA-ES [CFFS10] and Riemannian Particle Swarm Optimization [BIA10, BA10] on the $n$-sphere, Grassmannian manifolds, and Jacob's ladder respectively. Jacob's ladder, in particular, is a manifold of potentially infinite genus and cannot be addressed by traditional (constraint) optimization techniques on Euclidean spaces, which necessitates the development of manifold optimization algorithms.

## References

[ABG07] P-A Absil, Christopher G Baker, and Kyle A Gallivan. Trust-region methods on Riemannian manifolds. *Foundations of Computational Mathematics*, 7(3):303–330, 2007.

[AC10] Shun-ichi Amari and Andrzej Cichocki. Information geometry of divergence functions. *Bulletin of the Polish Academy of Sciences: Technical Sciences*, 58(1):183–195, 2010.

[AM97] Uwe Abresch and Wolfgang T Meyer. Injectivity radius estimates and sphere theorems. *Comparison geometry*, 30:1–47, 1997.

[Ama98] Shun-Ichi Amari. Natural gradient works efficiently in learning. *Neural computation*, 10(2):251–276, 1998.

[AMS09] P-A Absil, Robert Mahony, and Rodolphe Sepulchre. *Optimization algorithms on matrix manifolds*. Princeton University Press, 2009.

[AN00] S. Amari and H Nagaoka. *Methods of Information Geometry*, volume 191 of *Translations of Mathematical monographs*. Oxford University Press, 2000.

[BA10] Pierre B Borckmans and Pierre-Antoine Absil. Oriented bounding box computation using particle swarm optimization. In *ESANN*, 2010.

[Bey14] Hans-Georg Beyer. Convergence analysis of evolutionary algorithms that are based on the paradigm of information geometry. *Evolutionary Computation*, 22(4):679–709, 2014.

[BIA10] Pierre B Borckmans, Mariya Ishteva, and Pierre-Antoine Absil. A modified particle swarm optimization algorithm for the best low multilinear rank approximation of higher-order tensors. In *International Conference on Swarm Intelligence*, pages 13–23. Springer, 2010.

[CEE75] Jeff Cheeger, David G Ebin, and David Gregory Ebin. *Comparison theorems in Riemannian geometry*, volume 9. North-Holland Publishing Company Amsterdam, 1975.

[CFFS10] Sebastian Colutto, Florian Fruhauf, Matthias Fuchs, and Otmar Scherzer. The cma-es on riemannian manifolds to reconstruct shapes in 3-d voxel images. *IEEE Transactions on Evolutionary Computation*, 14(2):227–245, 2010.

[CM87] Paul H Calamai and Jorge J Moré. Projected gradient methods for linearly constrained problems. *Mathematical programming*, 39(1):93–116, 1987.

[HK04] Nikolaus Hansen and Stefan Kern. Evaluating the cma evolution strategy on multimodal test functions. In *International Conference on Parallel Problem Solving from Nature*, pages 282–291. Springer, 2004.

[Kli61] Wilhelm Klingenberg. Über riemannsche mannigfaltigkeiten mit positiver krümmung. *Commentarii Mathematici Helvetici*, 35(1):47–54, 1961.

[Leb02] Guy Lebanon. Learning riemannian metrics. In *Proceedings of the Nineteenth conference on Uncertainty in Artificial Intelligence*, pages 362–369. Morgan Kaufmann Publishers Inc., 2002.

[Lee01] John M Lee. *Introduction to smooth manifolds*. Springer, 2001.

[Lee06] John M Lee. *Riemannian manifolds: an introduction to curvature*, volume 176. Springer Science & Business Media, 2006.

[MP15] Luigi Malagò and Giovanni Pistone. Gradient flow of the stochastic relaxation on a generic exponential family. In *AIP Conference Proceedings*, volume 1641, pages 353–360. AIP, 2015.

[Pen04] Xavier Pennec. *Probabilities and statistics on Riemannian manifolds: A geometric approach*. PhD thesis, INRIA, 2004.

[Pen06] Xavier Pennec. Intrinsic statistics on riemannian manifolds: Basic tools for geometric measurements. *Journal of Mathematical Imaging and Vision*, 25(1):127, 2006.

[R+64] Walter Rudin et al. *Principles of mathematical analysis*, volume 3. McGraw-hill New York, 1964.

[WSPS08a] Daan Wierstra, Tom Schaul, Jan Peters, and Juergen Schmidhuber. Natural evolution strategies. In *2008 IEEE Congress on Evolutionary Computation (IEEE World Congress on Computational Intelligence)*, pages 3381–3387. IEEE, 2008.

[WSPS08b] Daan Wierstra, Tom Schaul, Jan Peters, and Jürgen Schmidhuber. Fitness expectation maximization. In *International Conference on Parallel Problem Solving from Nature*, pages 337–346. Springer, 2008.

[ZM04] Qingfu Zhang and Heinz Muhlenbein. On the convergence of a class of estimation of distribution algorithms. *IEEE Transactions on evolutionary computation*, 8(2):127–136, 2004.

# Chapter 9
# Illustrative Examples

**Abstract** This chapter consisits of several illustrative examples comparing Extended RSDFO with state-of-the-art manifold optimization algorithms such as Riemannian Trust-Region method, Riemannian CMA-ES and Riemannian Particle Swarm Optimization on the $n$-sphere, Grassmannian manifold, and Jacob's ladder. Jacob's ladder, in particular, is a non-compact manifold of countably infinite genus, which cannot be expressed as polynomial constraints and does not have a global representation in an ambient Euclidean space. Optimization problems on Jacob's ladder therefore cannot be addressed by traditional (constraint) optimization techniques on Euclidean spaces, which necessitates the development of manifold optimization algorithms.

In this chapter we compare Extended RSDFO (Algorithm 8.2, Sect. 8.2.2) against three state-of-the-art manifold optimization algorithms from the literatures, belonging to three distinctive categories, reviewed in Chap. 7:

1. **[Gradient-based optimization]**: Manifold Trust-Region method (RTR, Sect. 7.2.1) [ABG07, AMS09]
2. **[Stochastic optimization]**: Riemannian CMA-ES (RCMA-ES, Sect. 7.2.2) [CFFS10], and
3. **[Population-based optimization]** Riemannian Particle Swarm Optimization (R-PSO, Sect. 7.2.3) [BIA10, BA10].

We show that Extended RSDFO is either comparable to or out-performs the rival algorithms on both the local and global scape on a set of multi-modal optimization problems over a variety of search space manifolds. The experiments in this chapter are implemented with ManOpt version 4.0 [BMAS14] in MATLAB R2018a, and the remainder of the chapter is organized as follows:

1. In Sect. 9.1, we discuss the local restrictions and implementation assumptions of the manifold optimization algorithms in the literature, and discuss why they are not necessary for Extended RSDFO.
2. From Sects. 9.2 to 9.5, we compare Extended RSDFO against the aforementioned rival algorithms. In Sect. 9.2, we describe the hyperparameters and the setup of the algorithms involved in the experiments. The comparisons of the optimization



R. S. Fong and P. Tino, *Population-Based Optimization on Riemannian Manifolds*, Studies in Computational Intelligence 1046,
https://doi.org/10.1007/978-3-031-04293-5_9

methods are done on **minimization** problems over lower dimensional search space manifold, such as:

- $S^2$, the unit 2-sphere in Sect. 9.3
- Grassmannian manifolds $Gr_{\mathbb{R}}(p, n)$ in Sect. 9.4, and
- Jacob's ladder in Sect. 9.5.

The synthetic experiment on Jaccob's ladder, in particular, is introduced to motivate and necessitate the development of manifold optimization methods. Jacob's ladder is a non-compact manifold of countable infinite genus, which does not have a global representation in an ambient Euclidean space and cannot be addressed by traditional (constraint) optimization techniques on Euclidean spaces. Expected fitness in the experiments are approximated by Monte Carlo sampling with respect to their corresponding probability distributions.

3. In Sect. 9.6, we conclude the chapter by discussing the experimental results and their implications in detail.

## 9.1 On the Assumptions of Manifold Optimization Algorithms in the Literature

In Sect. 7.1, we discussed how manifold optimization methods in the literature and the statistical approach (Sect. 4.2) of constructing local probability distributions on manifolds both made the same assumptions on the base manifold. That is, in order to avoid imposing the locality on the optimization process and statistical estimations, the base manifold is assumed to be **complete**. This assumption in turn restricts the generality of both the optimization algorithms and the statistical approach of constructing local probability distributions, as most manifolds are *not complete*. In this section we flesh out the assumptions of each of the rival manifold optimization methods to be compared against Extended RSDFO in this chapter, and describe how the assumptions of the rival algorithms are *not* necessary for Extended RSDFO.

All three rival algorithms described in Chap. 7, RTR (Sect. 7.2.1, [ABG07]), Riemannian CMA-ES (Sect. 7.2.3, [CFFS10]), and Riemannian PSO (Sect. 7.2.2, [BIA10]), employ the assumption that the search space manifold $M$ is complete. This assumption equivalently means the Riemannian exponential map is defined on the entire tangent space of any point on $M$. Given a tangent vector (search direction) of an arbitrary length, the completeness of $M$ allows the algorithms to obtain new search iterates by "projecting" the current search iterate along any such direction for time 1. In particular, if the upper-bound of step length is set to be larger than the injectivity radius of $M$, then the algorithms can generate search iterates that escapes the current normal neighbourhood.

In Extended RSDFO, this assumption that $M$ is complete is not necessary to overcome the confines of a single normal neighbourhood. By the discussions in Chap. 8, the optimization process of Extended RSDFO is guided by an overarching family of mixture densities on an expanding totally bounded subset of $M$. This allows Extended RSDFO to utilize search information across a growing set of geodesic balls, overcoming the locality of single normal neighbourhood without losing generality in assuming completeness of $M$.

In addition to the completeness of the search space manifold $M$, Riemannian PSO requires an even stronger assumption to operate properly: the Riemannian logarithm map has to be defined between *any* two points; this is generally impossible. The Riemannian logarithm map is the inverse of Riemannian exponential map, which *only* exists in the region where the Riemannian exponential map is a diffeomorphism. In other words, the Riemannian logarithm map is only defined within the normal neighbourhood, and undefined othewise. Therefore in the Riemannian adaptation of PSO described in [BIA10], the full PSO step is only possible if all search agents are within *one single* normal neighourhood. This assumption is not necessary for Extended RSDFO, RTR, and RCMA-ES.

For "smaller" manifolds such as $n$-spheres and Grassmannian manifolds, the Riemannian logarithm map between (almost) any two points is well-defined due to a "sufficiently global parametrization" or embedding from an ambient Euclidean space.[1] However, for "larger" manifolds such as Jacob's ladder, some components of the Riemannian PSO step will be impossible, as we shall discuss in Sect. 9.5.

## 9.2 Hyperparameters of Algorithms and Experiment Set-Up

In this section we describe the hyperparameters and the set-up of the algorithms used in the subsequent sections. The parameters of the manifold optimization algorithms in the literature are chosen according to their original papers ((RTR) [AMS09], (RCMA-ES) [CFFS10], and (R-PSO) [BIA10]) as the dimensions of the experiments in this book are similar the ones described in their original papers.

***Setup of Extended RSDFO*** (Sect. 8): the RSDFO core is given by the generic framework described in Algorithm 8.1. Let $M$ denote the Riemannian manifold search space. The families of parametrized probability distributions $S_x$ on each tangent space $T_xM$ are set to be multivariate Gaussian densities centered at $\mathbf{0} \subset T_xM$ restricted to $B(\mathbf{0}, \text{inj}(x)) \subset U_x \subset T_xM$ (properly renormalized), denoted by $N(\cdot|\mathbf{0}, \Sigma)$. The initial statistical parameter is set to be $(\mathbf{0}, \Sigma) = \left(\mathbf{0}, \text{Id}_{\dim(M)}\right)$ (spherical Gaussian), and the covariance matrix is iteratively re-estimated throughout the execution (see Algorithm 8.1. Each iteration of local RSDFO is given a fixed budget of sample "parents" and the estimation is based on a fixed number of "off-springs". The local RSDFO algorithms are terminated if the improvement of solutions is lower than the threshold $10^{-14}$. Extended RSDFO is terminated when all local RSDFO algorithms around the current centroids terminate (see Sect. 8.2.7). The non-increasing non-zero function $\tau(k)$ modulating the amount of exploration on the $k$th iteration of

[1] For Grassmannian manifolds see [AMS04].

**Table 9.1** Parameters of RTR

| Parameter | Value | Meaning |
|---|---|---|
| $\overline{\Delta}$ | $\pi$ | Upperbound on step length |
| $\Delta_0$ | $\frac{\overline{\Delta}}{8}$ | Initial step length |
| $\rho'$ | 0.1 | Parameter that determines whether the new solution is accepted |

Extended RSDFO (see Sect. 8.2.3) is given by:

$$\tau(k) = \frac{6}{10} \cdot e^{-(0.015)\cdot k} \quad .$$

Furthermore, for the sake of computational efficiency, the exploration distribution $U = \text{unif}\left(\partial\left(W^k\right)\right)$ in line 2 of Extended RSDFO (see Sect. 8.2.6) will be simplified as follows: For each step we pick 5 explored centroids uniformly randomly. For each chosen centroids, 50 points are generated on the *sphere* of injectivity radius: $\exp_{x_\alpha}\left(B(\mathbf{0}, \text{inj}(x_\alpha))\right)$.[2] If a sufficient number of boundary points is accepted (according to Eq. (8.16)), then the appropriate number of boundary points will be added to the current centroids. However, if all the sampled points are rejected, then Extended RSDFO will not sample new boundary points. The algorithm will proceed on with the current set of centroids.

***Setup of RTR*** (Sect. 7.2.1, [ABG07]): The parameters of RTR are inherited from the classical Euclidean version of Trust Region [NW06], and are set to be their default values in ManOpt's implementation summarized in Table 9.1.

***Setup of RCMA-ES*** (Sect. 7.2.3, [CFFS10]) For RCMA-ES, the parameters of inherited directly from the Euclidean version [Han06]. The parameters in the following experiments are set to be the default parameters described in [CFFS10]. Let $N$ denote the dimension of the search space manifold, the parameters of RCMA-ES are summarized in Table 9.2.

***Setup of R-PSO*** (Sect. 7.2.2, [BIA10]): Finally, the parameters of R-PSO are set to be the default parameters in ManOpt [BMAS14] and the function of inertial coefficient is defined according to [BIA10].

$$\begin{aligned} &\text{nostalgia coefficient} = 1.4, \quad \text{social coefficient} = 1.4, \\ &\text{inertia coefficient} \ = \text{monotonic decreasing linear function from 0.9 to 0.4} \quad . \end{aligned}$$

[2] Equivalently, this means we set $j_{x_\alpha} = \text{inj}(x_\alpha)$ in Sect. 8.2.6.

**Table 9.2** Parameters of RCMA-ES described in [CFFS10]

| Parameter | Value | Meaning |
|---|---|---|
| $m_1$ | Depends on the manifold | Number of sample parents per iteration |
| $m_2$ | $\frac{m_1}{4}$ | Number of offsprings. |
| $w_i$ | $\log\left(\frac{m_2+1}{i}\right) \cdot \left(\sum_{j=1}^{m_2} \log\left(\frac{m_2+1}{j}\right)\right)^{-1}$ | Recombination coefficient |
| $m_{\text{eff}}$ | $\left(\sum_{i=1}^{m_2} w_i^2\right)^{-1}$ | Effective samples |
| $c_c$ | $\frac{4}{N+4}$ | Learning rate of anisotropic evolution path |
| $c_\sigma$ | $\frac{m_{\text{eff}}+2}{N+m_{\text{eff}}+3}$ | Learning rate of isotropic evolution path |
| $\mu_{\text{cov}}$ | $m_{\text{eff}}$ | Factor of rank-$m_2$-update of Covariance matrix |
| $c_{\text{cov}}$ | $\frac{2}{\mu_{\text{cov}}\left(N+\sqrt{2}\right)^2} + \left(1 - \frac{1}{\mu_{\text{cov}}}\right) \min\left(1, \frac{2\mu_{\text{cov}}-1}{(N+2)^2+\mu_{\text{cov}}}\right)$ | Learning rate of covariance matrix update |
| $d_\sigma =$ | $1 + 2\max\left(0, \sqrt{\frac{m_{\text{eff}}-1}{N+1}}\right) + c_\sigma$ | Damping parameter |

## 9.3 Sphere $S^2$

The first experiment considers the objective function discussed in [CFFS10] on $S^2$ with the spherical coordinates $(\theta, \phi)$:

$$f : S^2 \to \mathbb{R}$$
$$f(\theta, \phi) = -2\cos(6 \cdot \theta) + 2\cos(12 \cdot \phi) + \left(\frac{\pi}{12} - \phi\right)^2 + \frac{7}{2}\theta^2 + 4 \quad .$$

$f$ is a multi-modal problem with global minimum located at $(\theta^*, \phi^*) = \left(0, \frac{\pi}{12}\right)$ with objective value 0. The heat map of the objective function is illustrated in Fig. 9.1.

Injectivity radius on $S^2$ is given by $\text{inj}(x) = \pi$ for all $x \in S^2$[TAV13], and the Riemannian exponential map of $S^2$ is given by [AMS09]:

$$\exp_x(v) = \gamma_v(t)|_{t=1} = \cos(||v|| \cdot t)\, x + \sin(||v|| \cdot t) \left.\frac{v}{||v||}\right|_{t=1} \quad ,$$

where $\gamma_v(t)$ is the geodesic on $S^2$ starting at $x$ with initial velocity $v \in T_x S^2$ . For the following experiments, the additional parameters of Extended RSDFO are given by:

$$N_{random} = 2, \quad N_{cull} = 2 \quad .$$

Each optimization algorithm is given a budget of 10000 function evaluations. Extended RSDFO initializes with 2 initial centroids $X^0$ randomly generated on the

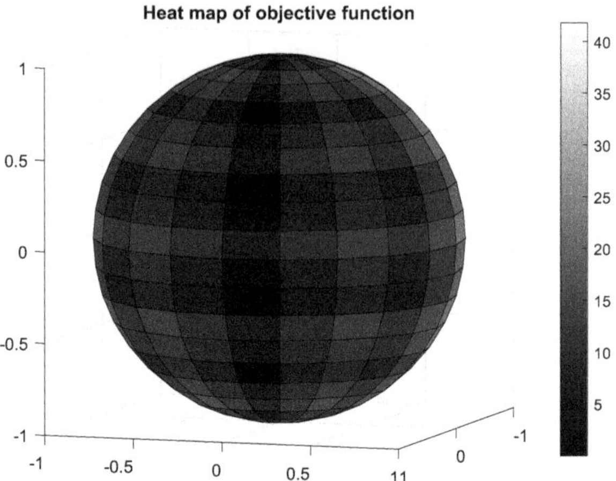


**Fig. 9.1** Heat map of $f$ on $S^2$

sphere $S^2$. The local RSDFO core is given a budget of 50 "parents", and the parameter estimation will be based on the best 10 "offsprings". RTR and RCMA-ES is initialized with one centroid randomly chosen from the same set of initial centroids $X^0$. RCMA-ES is given a budget of $m_1 = 40$ parents per iteration. R-PSO initialized with 10 copies of $X^0$, which allows 20 "agents" to perform 20 function evaluations per iteration. The expected fitness of the stochastic algorithms are approximated using 10 Monte Carlo samples with respect to their corresponding probability distributions. Each algorithm is repeated 200 times with the parameters described in the beginning of this section. The gradient and the Hessian of the objective function is not provided and each estimation of the gradient or the Hessian in RTR counts as a single function evaluation respectively, even though in practice it may take up additional resources. The algorithms (and the RSDFO steams in Extended RSDFO) terminate when they converge locally or when they are sufficiently ($10^{-6}$) close to the global optimum with objective value 0.

Furthermore, we perform an additional set of experiments with a slightly relaxed exploration distribution described in Sect. 8.2.6. In particular, we choose non-zero real number $1 > \epsilon_b > 0$, such that for each iteration $k > 0$ the "exploration" centroids (described in Sect. 8.2.6) is generated from $\partial\left(\bigcup_{\alpha \in \Lambda_A^k} \tilde{B}_\alpha\right)$, with $\tilde{B}_\alpha := \exp_{x_\alpha}(B(\mathbf{0}, \epsilon_b \cdot \text{inj}(x_\alpha))) \subset \exp_{x_\alpha}(B(\mathbf{0}, \text{inj}(x_\alpha)))$.

The result of 200 experimental runs is summarized in Table 9.3. The second and third column of the Table 9.3 shows the number of local minimum and global minimum attained by the optimization methods within the 200 experiments respectively.

From Table 9.3, we observe that Extended RSDFO with a basic RSDFO core (Algorithm 8.1) performs better than RCMA-ES even on the 2-sphere. While R-PSO out-performs the rest of the rival algorithms slightly at the expense of higher number of objective function evaluations. The slightly relaxed version of Extended RSDFO with $\epsilon_b = 0.4$ is closely comparable to R-PSO using less than half the computational resources. We will discuss the experimental results in further detail in Sect. 9.6.1.

**Table 9.3** Experimental results on $S^2$ with 200 runs

| Method | No. of local min | No. of global min | Avg. $f$ eval |
|---|---|---|---|
| Ext. RSDFO with $\epsilon_b = 1$ | 72 | 128 | 5154 |
| RCMA-ES | 104 | 96 | 5518.6 |
| R-PSO | 33 | 167 | 10000 |
| RTR | 192 | 8 | 3093.9 |
| Ext. RSDFO with $\epsilon_b = 0.4$ | 46 | 154 | 4648 |

Experimental results of a sample run is provided in Fig. 9.2a. In this experiment both versions of Extended RSDFO and RCMA-ES attained the global minimum, whilst R-PSO and RTR are stuck in local minima. The search centroids of model-based algorithms (Extended RSDFO and RCMA-ES) and sampled points of R-PSO are presented in Fig. 9.2b, c respectively. We observe that the movement of search centroids of model-based algorithms (Extended RSDFO and RCMA-ES) exhibit varing but converging sequence towards the global optima. On the other hand, the search agents of R-PSO gravitates towards the local min, similar to its Euclidean counterpart.

## 9.4 Grassmannian Manifolds

For experiments on Grassmannian manifolds $Gr_{\mathbb{R}}(p, n)$ with $(p, n) = (2, 4), (2, 5)$, we consider a composition of Gramacy-Lee Function [AA19, GL12], and the Riemannian distance function $d_{Gr}$ with respect to $I_{n\times p} \in Gr_{\mathbb{R}}(p, n)$ :

$$
\begin{aligned}
f &: Gr_{\mathbb{R}}(p, n) \to \mathbb{R} \\
f(x) &= \frac{\sin\left(10\pi \cdot d_{Gr}\left(x, I_{n\times p}\right)\right)}{2 \cdot d_{Gr}\left(x, I_{n\times p}\right)} + \left(d_{Gr}\left(x, I_{n\times p}\right) - 1\right)^4 \quad ,
\end{aligned}
$$

where $d_{Gr}\left(\cdot, I_{n\times p}\right)$ denote the Riemannian distance function on $Gr_{\mathbb{R}}(p, n)$ with respect to $I_{n\times p}$. The objective function is illustrated in Fig. 9.3, notice the $x$-axis is given by $d_{Gr}\left(x, I_{n\times p}\right)$.

The injectivity radius of Grassmannian manifolds $Gr_{\mathbb{R}}(p, n)$ is $\frac{\pi}{2}$ for all $x \in Gr_{\mathbb{R}}(p, n)$ [Prá96, TAV13], and the Riemannian exponential map of is given by [AMS04]:

$$
\exp_x(v) = \gamma_v(t)\big|_{t=1} = \cos\left(||v|| \cdot t\right) x + \sin\left(||v|| \cdot t\right) \left.\frac{v}{||v||}\right|_{t=1} \quad ,
$$

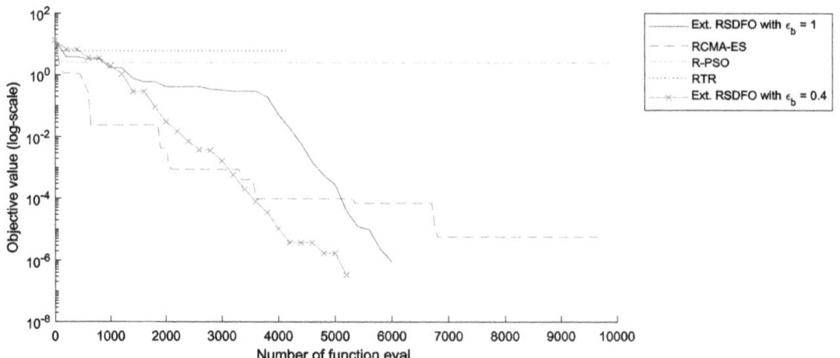


(a) A sample comparison of quality of solution obtained by each algorithm.

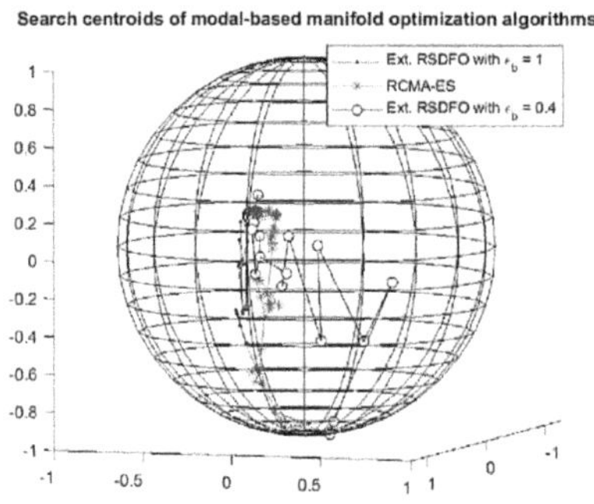


(b) Movement of search centroids in model-based algorithms. Notice that even the algorithms share the same initial search centroid position at the bottom of the sphere, their search paths differ as they depends on the estimated statistical model.

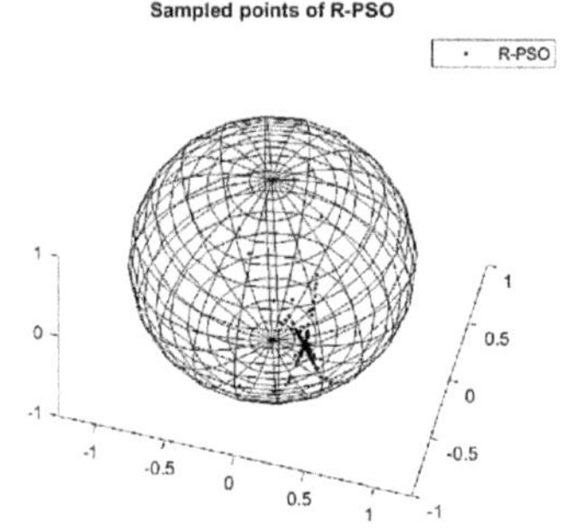


(c) Search agents generated by R-PSO.

**Fig. 9.2** Experimental results of one of the experiments on $S^2$

where $\gamma_v(t)$ is the geodesic on $Gr_{\mathbb{R}}(p, n)$ starting at $x$ with initial velocity $v \in T_x Gr_{\mathbb{R}}(p, n)$. For the following experiment, the additional parameters of Extended RSDFO are given by:

$$N_{random} = 2, \quad N_{cull} = 2 \quad .$$

The initialization of the following experiments is similar to that of $S^2$ discussed in the previous section, the specification is included for the sake of completeness: Extended RSDFO initializes with 2 initial centroids $X^0$ randomly generated on the $Gr_{\mathbb{R}}(p, n)$. The local RSDFO core is given a budget of 120 and 200 "parents" with parameter estimation based on the best 40, 50 "offsprings" on $Gr_{\mathbb{R}}(2, 4)$ and $Gr_{\mathbb{R}}(2, 5)$ respectively. RTR and RCMA-ES is initialized with one centroid randomly chosen from

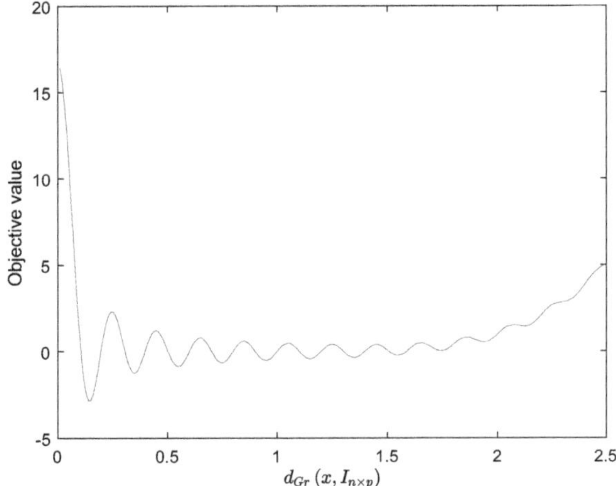


**Fig. 9.3** Illustration of objective function $f$ on $Gr_{\mathbb{R}}(p, n)$, notice the $x$-axis is given by the Riemannian distance $d_{Gr}$ of $x$ relative to $I_{n\times p}$

$X^0$. RCMA-ES is given a budget of $m_1 = 80$ parents per iteration. R-PSO initialized with 20 copies of $X^0$, which allows 40 "agents" performing 40 function evaluations per iteration. Each algorithm is repeated 100 times with the parameters described in the beginning of this section. The gradient and the Hessian of the objective function is not provided. The expected fitness of the stochastic algorithms are approximated using 40,50 Monte Carlo samples with respect to their corresponding probability distributions on $Gr_{\mathbb{R}}(2, 4)$ and $Gr_{\mathbb{R}}(2, 5)$ respectively. Each estimation of the gradient or the Hessian in RTR counts as a single function evaluation respectively.

Furthermore, we once again perform an additional set of experiments with a slightly relaxed exploration distribution described in Sect. 8.2.6. Following the discussion in the previous section, we choose non-zero real number $1 > \epsilon_b > 0$, such that for each iteration $k > 0$ the "exploration" centroids (described in Sect. 8.2.6) is generated from $\partial\left(\bigcup_{\alpha\in\Lambda_A^k}\tilde{B}_\alpha\right)$, with $\tilde{B}_\alpha := \exp_{x_\alpha}(B(\mathbf{0}, \epsilon_b\cdot \mathrm{inj}(x_\alpha))) \subset \exp_{x_\alpha}(B(\mathbf{0}, \mathrm{inj}(x_\alpha)))$.

We perform two sets of experiments on two Grassmannian manifolds $Gr_{\mathbb{R}}(2, 4)$, $Gr_{\mathbb{R}}(2, 5)$, where each optimization algorithm is given a budget of 24000 and 40000 function evaluations on $Gr_{\mathbb{R}}(2, 4)$ and $Gr_{\mathbb{R}}(2, 5)$respectively. Due to the complexity of the problem, the convergence criteria is slightly relaxed: the algorithms and the RSDFO steams in Extended RSDFO are said to converge globally if they are sufficiently close (with threshold $10^{-8}$) to the global optimum with objective value $-2.87$, meaning the search iterate is within the basin of the global optimum.

The result of 100 experimental runs on $Gr_{\mathbb{R}}(2, 4)$ and $Gr_{\mathbb{R}}(2, 5)$ is summarized in Tables 9.4 and 9.5 respectively below. The second and third column of the tables shows the number of local minimum and global minimum attained by the optimization methods within the 100 experiments respectively.

From Tables 9.4 and 9.5, we observe that both versions of Extended RSDFO are comparable to R-PSO, which out-performs the rival algorithms. While these three algorithms achieves a high success rate in determining the global solution

**Table 9.4** Experimental results on $Gr_{\mathbb{R}}(2, 4)$ with 100 runs

| Method | Local min | Global min | Avg. $f$ eval |
|---|---|---|---|
| Ext. RSDFO with $\epsilon_b = 1$ | 4 | 96 | 24000 |
| RCMA-ES | 61 | 39 | 9786.4 |
| R-PSO | 3 | 97 | 24000 |
| RTR | 100 | 0 | 22719.54 |
| Ext. RSDFO with $\epsilon_b = 0.5$ | 5 | 95 | 24000 |

**Table 9.5** Experimental results on $Gr_{\mathbb{R}}(2, 5)$ with 100 runs

| Method | Local min | Global min | Avg. $f$ eval |
|---|---|---|---|
| Ext. RSDFO with $\epsilon_b = 1$ | 70 | 30 | 40000 |
| RCMA-ES | 91 | 9 | 13271.2 |
| R-PSO | 44 | 56 | 40000 |
| RTR | 100 | 0 | 39445.47 |
| Ext. RSDFO with $\epsilon_b = 0.5$ | 59 | 41 | 40000 |

of the experiments on $Gr_{\mathbb{R}}(2, 4)$, the success rates of the algorithms dropped as the dimension increases on $Gr_{\mathbb{R}}(2, 5)$. We will discuss the experimental results in further detail in Sect. 9.6.1.

Experimental results of a sample run on $Gr_{\mathbb{R}}(2, 4)$ and $Gr_{\mathbb{R}}(2, 5)$ is provided in Fig. 9.4a, b respectively below. In the experiment on $Gr_{\mathbb{R}}(2, 4)$, all algorithms *except* RTR and RCMA-ES attained the global minimum. The plot shows how the convergence rate of the algorithms vary; Extended RSDFO takes more steps to explore the manifold before converging to an optimum. In the experiment on $Gr_{\mathbb{R}}(2, 5)$, both versions of Extended RSDFO attained the global optimum, while the other algorithms converged to local optima. It is worth noting that, whilst RCMA-ES convergences sharply, it is prone to be stuck in local minima at higher dimensions.

## 9.5 Jacob's Ladder

In this section we discuss the experimental results of an optimization problem on a representation of **Jacob's ladder** surface – a manifold of infinite genus [Ghy95]. Jacob's ladder will be represented by the connected sum of countable number of tori [Spi79], illustrated in Fig. 9.5.[3]

[3] The general geometrical structure of manifolds of infinite genus is beyond the scope of this book. Interested readers are referred to [PS81, FKT95] and the relevant publications.

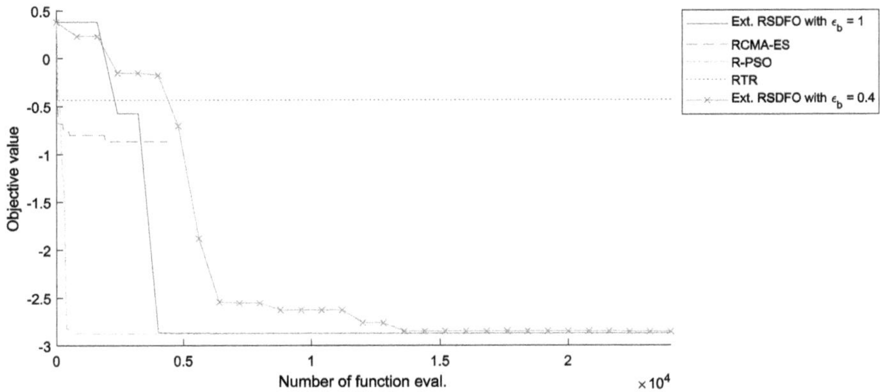


(a) Comparison of quality of solution obtained by each algorithm in $Gr_{\mathbb{R}}(2, 4)$.

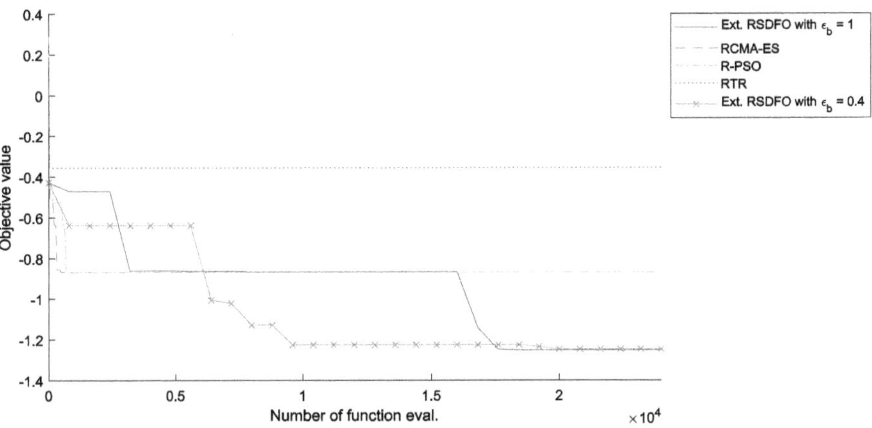


(b) Comparison of quality of solution obtained by each algorithm in $Gr_{\mathbb{R}}(2, 5)$.

**Fig. 9.4** Experimental results of one of the experiments on $Gr_{\mathbb{R}}(p, n)$

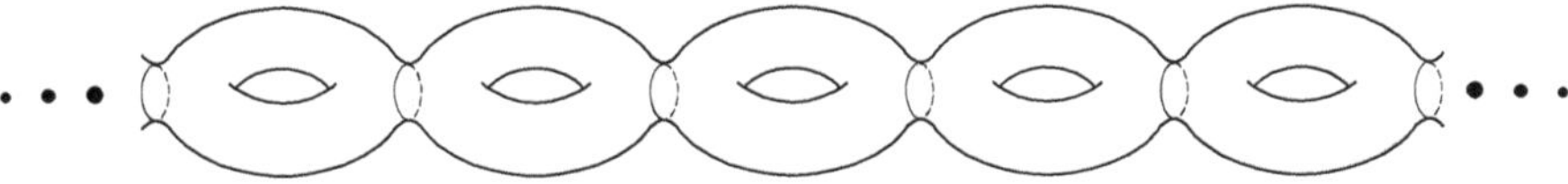

**Fig. 9.5** Illustration of Jacob's ladder surface as an infinite connected sum of tori

Traditionally, manifold optimization falls under the framework of constraint optimization over Euclidean spaces. Indeed, due to Whitney's embedding theorem [Whi44a, Whi44b], all manifolds can be embedded in a sufficiently large ambient Euclidean space. As traditional optimization techniques are more established and well-studied in Euclidean spaces, one would be more inclined to address optimization problems on Riemannian manifold by first finding an embedding onto the manifold, and then applying familar classical optimization techniques.

Riemannian manifolds discussed in the manifold optimization literature [AMS09, AH19] are mostly compact manifolds such as $n$-spheres, Steifel manifolds, or Grassmannian manifolds. These manifolds all have global vectorial (matrix) representations, which provide a global parametrization of the manifold. Moreover, these manifolds are all compact manifolds, which can be expressed as zero set of polynomials also known as real affine algebraic varieties [Tog73, BK89]. This means compact manifolds can be written as polynomial constraints in an ambient Euclidean space.

To motivate and necessitate the use of intrinsic Riemannian manifold optimization, we present an optimization problem on Jacob's ladder as an example of a Riemannian manifold search space that does not have a global parametrization. Furthermore, Jacob's ladder is a surface of countably infinite genus, meaning it cannot be expressed as polynomial constraints in the manifold.[4] It is therefore difficult to formulate optimization problems on Jacob's ladder as a constraint optimization problem.

For the rest of the discussion and in the following experiment we assume no upperbound on the number of tori in the Jacob's ladder chain. A priori the optimization algorithms does *not* know the size of the tori chain and so the manifold in the optimization problem cannot be formulated as constraints beforehand. In this situation, the only natural approach is to use intrinsic Riemannian manifold optimization methods that "crawl" on the manifold.

Jacob's ladder represented by an infinite connected sum of tori is a non-compact complete Riemannian manifold, as the connected sum of complete manifolds is also complete. The local geometry of each torus in the Jacob's ladder is given by $S^1_R \times S^1_r$, where $R > r$ and $S^1_R$ and $S^1_r$ denote the major and minor circles of radius $R$ and $r$ respectively. A torus can be locally parametrized by the angles $(\theta, \varphi)$, where $\theta$ and $\varphi$ denote the angle of the minor circle and major circle respectively. The local exponential map at a point $x$ on a torus can be decomposed as $\left(exp_x^{S^1_R}, exp_x^{S^1_r}\right)$, this is due to the direct sum nature of the product Riemannian metric described in Sect. 6.4. If the geodesic ball of injectivity radius around a point $x$ intersects two tori in the Jacob's ladder, then the exponential map follows the transition function described below.

Let $T_1$ and $T_2$ denote two tori of the Jacob's ladder, and let $\psi_1$, $\psi_2$ denote smooth coordinate functions of $T_1$ and $T_2$ respectively. On the gluing part of the adjunction space, the transition map $\psi_2 \circ \psi_1^{-1}$ is given by the identity map. This is illustrated in Fig. 9.6.

[4] The Euler characteristic $\chi$ of a manifold can be express as the alternating sum of Betti numbers: $\chi = b_0 - b_1 + b_2 - \ldots$, where $b_i$ denote the $i$th Betti number. In the case of Jacob's ladder we have $b_0 = 1 = b_2$, and $b_j = 0$ for $j > 3$ due to the dimension of Jacob's ladder. Let $g$ denote the genus of the surface, then the Euler characteristic is given by $\chi = 2 - 2 \cdot g$. By combining the two equations we obtain $b_1 = 2 \cdot g$. Furthermore, the (sum of) Betti number of algebraic varieties is bounded above [Mil64]:, which means a manifold of infinite genus cannot be a real affine algebraic variety.

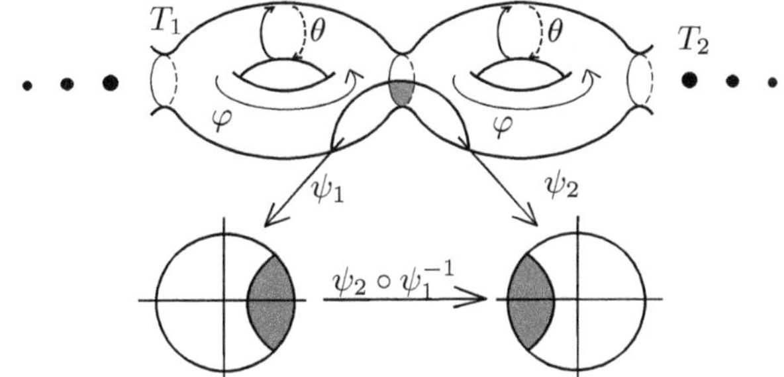


**Fig. 9.6** Illustration of minor and major rotational angles in the Jacob's ladder. The vertical smaller rotational angle represents the minor angle $\theta$, whereas the horizontal arc represents the major angle $\varphi$

Denote by $T_i$ the $i$th torus in Jacob's ladder $M$, described above. Due to the lack of global parametrization of $M$, we consider the **minimization** problem on $M$ with the following optimization function, composed of a "global" part and a "local part":

$$f : M \to \mathbb{R} \tag{9.1}$$

$$x \mapsto f_G(x) \cdot f_L(x) \quad , \tag{9.2}$$

where $f_G$ denote the "global" part of the objective function, which acts on the "torus number" of a point $x \in M$. To be precise, the function is given by:

$$f_G : M \to \mathbb{R}$$

$$n(x) \mapsto := \begin{cases} 0.05 & \text{if } n(x) = 0,\ 15, \textit{and} - 25 \\ \left[ -\left( \frac{\sin\left(\frac{4}{5}n(x)-15\right)}{\frac{4}{5}(n(x)-15)} \right)^2 + 1.05 \right] \cdot \left| \frac{n(x)}{20} \right|^{\frac{1}{10}} & \text{if } n(x) > \frac{15}{2} \\ \left[ -\left( \frac{\sin\left(\frac{4}{5}n(x)+25\right)}{\frac{4}{5}(n(x)+25)} \right)^2 + 1.05 \right] \cdot \left| \frac{n(x)}{20} \right|^{\frac{1}{10}} & \text{if } n(x) < -\frac{15}{2} \\ -\left( \frac{\sin\left(\frac{4}{5}n(x)\right)}{\frac{4}{5}n(x)} \right)^2 \cdot \left| \frac{n(x)}{20} \right|^{\frac{1}{10}} & \text{otherwise,} \end{cases}$$

where $n(x)$ is the integer denoting the numerical label of the torus in which the point $x \in M$ is located. The numerical label increases as we move towards the positive $x$-axis direction in the ambient Euclidean space and decreases as we move towards the negative $x$-axis direction. For instance, in the current implementation the torus centered at the origin is labelled torus number zero, and the next one towards the positive $x$-axis direction in the ambient Euclidean space is subsequently labelled 1. The function $f_G$ is illustrated in Fig. 9.7.

The local component of the objective function is given by Levy Function N.13 [SB17]:

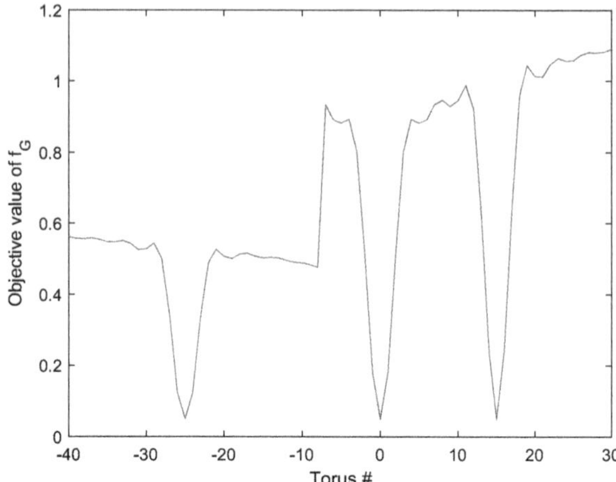


**Fig. 9.7** $f_G$: the "global" part of the objective function $f$ on Jacob's ladder

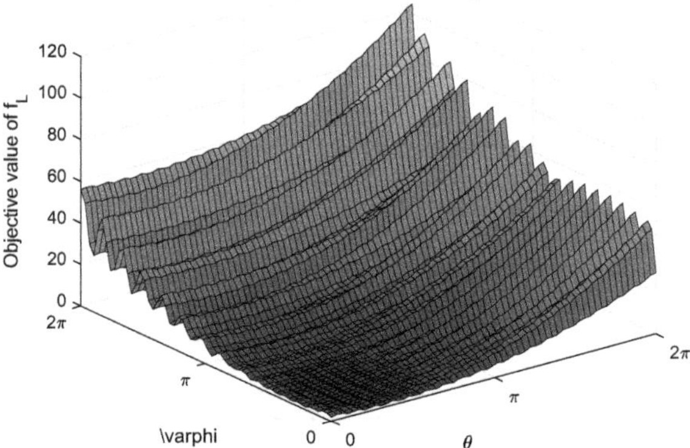


**Fig. 9.8** $f_L$: the "local" part of the objective function $f$ on Jacob's ladder

$$f_L : T_i \setminus \bigcup_{j \neq i} T_j \to \mathbb{R}$$

$$\theta, \varphi \mapsto \sin^2(3\pi\theta) + (\theta - 1)^2 \left(1 + \sin^2(3\pi\varphi)\right) + (\varphi - 1)^2 \left(1 + \sin^2(2\pi\varphi)\right) + 35 \quad .$$

The function $f_L$ is illustrated in Fig. 9.8 on the $(\theta, \varphi)$ axes. Levy N.13 is a non-convex, differentiable, multimodal function [AA19] with one global minimum at radian angles $(\theta^*, \varphi^*) = (1, 1)$ with objective value $f_L(\theta^*, \varphi^*) = 0 + 35 = 35$.

The "global-global" optima (the desired solution) is thus located in torus number $0, 15, -25$ with minor-major rotational angles $(\theta^*, \varphi^*) = (1, 1)$ with objective value $0.05 \cdot 35 = 1.75$. The torus numbered $0, 15, -25$ are therefore called "optimal" for the following experiments.

For the following experiment, the additional parameters of Extended RSDFO are given by:

$$N_{random} = 6, \quad N_{cull} = 3 \quad .$$

For each execution, Extended RSDFO initializes with a set of 5 initial centroids $X^0 := \{x_i^0\}_{i=1}^5$ randomly generated on the Jacob's ladder between torus number $-30$ to 30. The initial set of centroids $X^0$ is generated using a heuristic approach. RTR and RCMA-ES is initialized with one centroid randomly chosen from $X^0$. RCMA-ES is given a budget of $m_1 = 40$ parents per iteration. R-PSO initialized with 8 copies of $X^0$, which allows 40 "agents" performing 40 function evaluations per iteration. The expected fitness of the stochastic algorithms are approximated using 10 Monte Carlo samples with respect to their corresponding probability distributions. Each algorithm is repeated 100 times with the parameters described in the beginning of this section. For each execution, each algorithm is budgeted by $20,000$ function evaluations. The algorithms and the RSDFO steams in Extended RSDFO terminate when they converge locally or when they are sufficently close (with threshold $10^{-8}$)) to the optimal value at 1.75. The gradient and the Hessian of the objective function is not provided as we are treating it as a black-box optimization problem. Each estimation of the gradient or the Hessian in RTR counts as a single function evaluation respectively. The results of the experiments are summarized in Table 9.7. The meanings of the column labels of Table 9.7 are given in Table 9.6.

**Table 9.6** Detailed description of column labels in Table 9.7

| Column label | Meaning | Description |
|---|---|---|
| $\ell + \ell$ | Local-local optima | Algorithm converges without reaching optimal angles nor the optimal tori |
| $\ell + g$ | Local-global optima | Algorithm attains the optimal angles in a local torus **but** does not reach a global optimal torus |
| $g + \ell$ | Global-local optima | Algorithm reaches a global optimal torus **but** fails to find the local optimal angles within it |
| $g + g$ | Global-global optima (desired solution) | Algorithm converges to **both** an optimal torus and the optimal angles within the torus. This is the best possible outcome |
| avg. $f$ eval | Average function evaluation | The average function evaluation needed to either converge or reach the global optima |

Table 9.7 Experimental results on Jacob's ladder

| Method | $\ell+\ell$ | $\ell+g$ | $g+\ell$ | $g+g$ | avg. $f$ eval |
|---|---|---|---|---|---|
| Ext-RSDFO | 2 | 27 | 10 | 61 | 8483.8 |
| RCMA-ES | 1 | 76 | 0 | 23 | 4230.1 |
| RTR | 96 | 0 | 4 | 0 | 3086.2 |
| R-PSO | 2 | 5 | 56 | 37 | 20000 |

**Remark 9.1** It is important to note that, by the discussion in Sect. 9.1, two out of three components (nostalgia and social component) of the R-PSO step would be impossible without a Riemannian logarithm map between any two points on the manifold. Indeed, on Jacob's ladder the Riemannian logarithm map between two points on far away tori cannot be computed. For the following experiments, since the true Riemannian logarithm map is either difficult or impossible compute, we provide R-PSO with a heuristic version of the Riemannian logarithm map on the Jacob's ladder for the sake of comparison. In particular we allow "communication" between points beyond each other's normal neighbourhood.

To be precise, we provide a heuristic Riemannian logarithm map between points that are either within the same torus, or if the points are "close enough" (difference of torus number is 1). If two points are within the same tori, we compute the Riemannian logarithm map using the product geometry of $S^1 \times S^1$. For points that are "close enough", we provide a "general direction" if they are not in the same torus. That is, we first embed the two points in an ambient Euclidean space, determine the direction of the torus of the target point from the original point, then project the vector $\pm\mathbf{1}$ onto the tangent space of the major circle of the original torus. Otherwise if the pair of points $x, y$ is too far apart, we set $\log_x(y) = \mathbf{0}$. This is a sufficiently good approximation to the true Riemannian logarithm map on Jacob's ladder. The experimental results is given in Table 9.7.

It is important to note that this provides an unfair advantage to R-PSO over the other algorithms, as the other algorithms do not have this additonal requirement (see Sect. 9.1). Moreover, this requires additional computational resources not recorded by function evaluations.

Experimental results of a sample run is provided in Fig. 9.9. In Fig. 9.9a, we observe that only Extended RSDFO attained the global-global solution. Figure 9.9b provides a close-up view of Fig. 9.9a, which illustrates the performance of Extended RSDFO and R-PSO towards the global-global optima in the experiment. We observe that Extended RSDFO attained the global-global solution with around 15000 function evaluations, while R-PSO was only able to find a global-local solution (optimal torus but not the optimal angle) with the given budget of 20000 function evaluations.

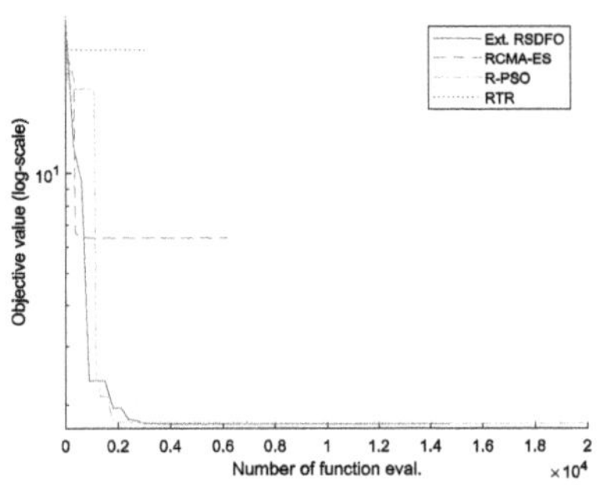


(a) Comparison of quality of solution obtained by each algorithm on Jacob's ladder

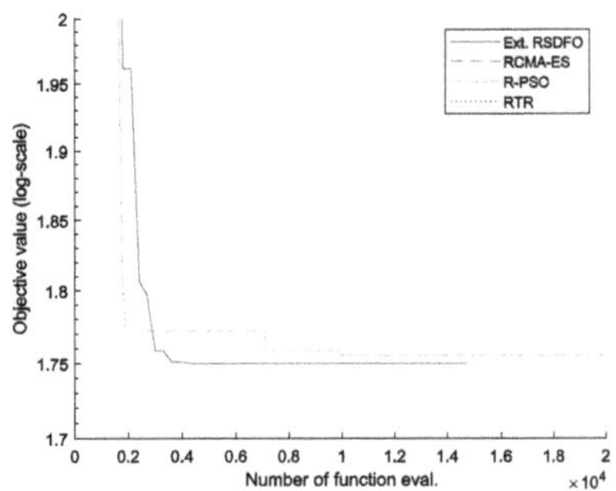


(b) Close-up of Figure 9.9a towards the global-global optima.

**Fig. 9.9** Experimental results of one of the experiments on Jacob's ladder

## 9.6 Discussion

From the experimental results of the previous sections, we observe that gradient-based line search methods (such as RTR) is less effective compared to population-based approaches such as model-based stochastic methods (Extended RSDFO, RCMA-ES) and meta-heuristic evolutionary strategies (R-PSO) when solving multi-modal optimization problems on Riemannian manifolds. We discuss the experiments in further detail in this section.

### *9.6.1 On Sphere and Grassmannian Manifold*

The Riemannian manifolds considered in the first two sets of experiments are the 2-sphere $S^2$ and low dimensional Grassmannian manifolds $Gr_{\mathbb{R}}(p, n)$. These two types of search space manifolds are "small", in the sense that they are low dimensional, compact, coverable by a small number of geodesic balls, and most importantly they admit a global parametrization (matrix representation).

The optimization problem on $S^2$ was from [CFFS10]. We observe that Extended RSDFO with the basic version of RSDFO core (Algorithm 8.1) performs better than RCMA-ES even on the sphere. On the other hand, R-PSO out-performs the rest of the algorithms, this is partly due to the abundance of computational resources (available computational budget relative to the size of the search manifold) and the amount of search agents available (relative to the size of the manifold).

The manifold $S^2$ is small, in the sense that it can be covered by geodesic balls centered around a handful of points. In particular, the implemented exploration distribution (described in the beginning of this section) often fails to find a boundary point before the point rejection threshold is reached after a few iterations. This resulted in less effective searches (in rejected points) and higher sensitivity to initial condition.

In light of this, we ran Extended RSDFO with additional flexibility in selecting new exploration centroids. By setting the geodesic balls to be $\epsilon_b = 0.4$ times the injectivity radius, we allow more overlapping between search regions generated by the search centroids, thus enhancing the exploration of search region. Indeed, the "relaxed version" of Extended RSDFO with $\epsilon_b = 0.4$ is less sensitive to initial conditions and has shown improvements over the version with $\epsilon_b = 1$. The relaxed version is also closely competitive with R-PSO in determing the global optima.

From the experiments on $Gr_{\mathbb{R}}(2, 4)$, we observe a similar behaviour as the experiment on $S^2$. For population-based stochastic algorithms: Extended RSDFO out performs RCMA-ES by searching with multiple centroids at the expense of additional resources. On the other hand both versions of Extended RSDFO are comparable to the population-based meta-heuristic R-PSO.

On $Gr_{\mathbb{R}}(2, 4)$ (a 4- dimensional manifold), the both versions of Extended RSDFO and R-PSO achieve a high success rate in determining the global solution. However, as we increase the dimension to $Gr_{\mathbb{R}}(2, 5)$ (a 6- dimensional manifold), the success rates of the algorithms dropped.

The drop in success rate for stochastic algorithms such as Extended RSDFO and RCMA-ES is partly due to the typical exponential increase in the amount of resources necessary to correctly estimate the underlying (local) stochastic modal. In particular, if we assume the amount of resources increases exponentially: for $Gr_{\mathbb{R}}(2, 4)$, the average function evaluation of Extended RSDFO is $24000 \approx (12.45)^4$, this means on $Gr_{\mathbb{R}}(2, 5)$ the expected function evaluation requirement would be approximately $(12.45)^6 \approx 3724055$, which is two orders of magnitude higher than our budget of 40000.

On the other hand, a manifold of higher dimension also means a larger amount of search centroids is needed to provide better exploration of the search space. This in turn affects the solution quality of both R-PSO and Extended RSDFO as the search agents remain unchanged.

From the two sets of experiments on $Gr_{\mathbb{R}}(2, 4)$ and $Gr_{\mathbb{R}}(2, 5)$ on, we observe that R-PSO is more scalable to the dimensional increases. On the other hand, model-based algorithms such as Extended RSDFO and RCMA-ES both require additional resources to estimate the correct model as the dimension increase.

### 9.6.2 On Jacob's Ladder

The optimization problem on Jacob's ladder described in Sect. 9.5 was deliberately set to be difficult: the objective function $f$ is multi-modal, non-convex, non-separable with three global optima. There are two notably challenging features of the objective function $f = f_G \cdot f_L$ from the "global" and "local" part respectively.

First of which comes from the "global" part of the objective function: $f_G$, which has a plateau for points with torus number within the range $[-19, -8]$. The aforemen-

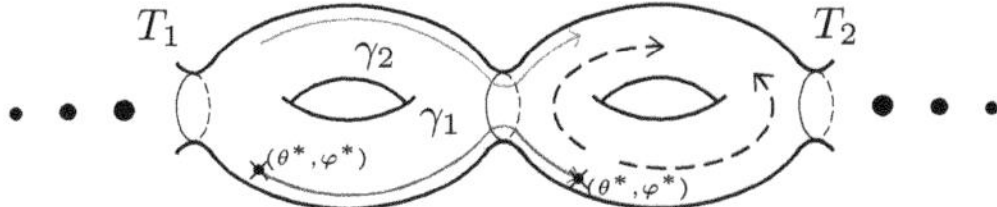


**Fig. 9.10** Illustration of the "trap" of $f_L$. The local optimal angles $(\theta^*, \varphi^*)$ are marked by the points labelled with $\times$. The objective value increases along dotted lines on $T_2$ (as well as $T_1$ and the other tori), as the "objective landscape", described by Fig. 9.8, "wraps around" the local coordinates of the tori. If an algorithm searches along $\gamma_2$ from $T_1$ to $T_2$, it may think $T_1$ is better. On the other hand, if an algorithm searches along $\gamma_1$, then $T_2$ may seem like the better torus

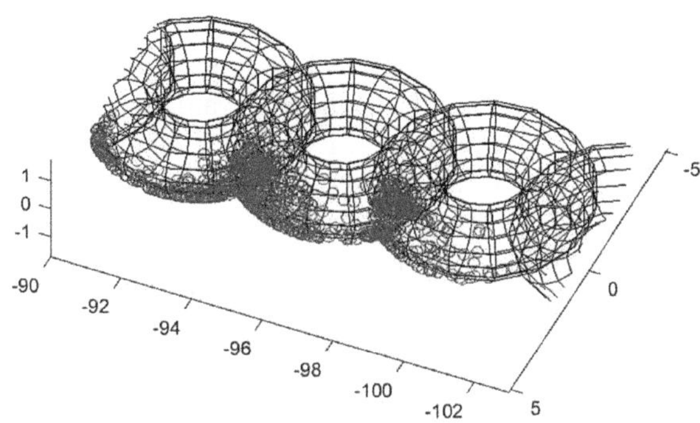


**Fig. 9.11** An illustration of path $\gamma_1$ of Fig. 9.10 from one of the experiments. This figure shows the final converging steps of Extended RSDFO. The search points generated from the converging distributions (from the RSDFO core) are concentrated around $(\theta^*, \varphi^*) = (1, 1)$ of each tori

tioned algorithms are prone to be stuck on centroids with tori number $-19 \ldots -8$, as $f_G(n_1) < f_G(n_2)$ for $n_1 \in [-19, -8]$ and $n_2 \in \{-3, -2, 2, 3, \ldots\}$. In particular, the search centroids within the plateau have slightly better function value than the other tori around the global optima, while being far away from global optima.

Secondly, the "local" part of the objective function $f_L$ also introduces complications due to the local parametrization of the tori. As illustrate in Figs. 9.10 and 9.11, as the search landscape of $f_L$ wraps around the tori, the solution estimate depends on how the search path travels between the tori in Jacob's ladder. The algorithms might wrongly estimate the overall objective function values of the neighboring torus depending on where the search path is.

From the experiments, we observe that Extended RSDFO achieves the highest global-global convergence rate amongst the manifold optimization algorithms in the experiments.

The current implementation of Extended RSDFO, with the basic version of SDFO (Algorithm 7.3) and a simplified version of exploration point selection (discussed in the beginning of this section), is competitive with R-PSO and RCMA-ES on the global and local scale respectively:

- **[Global scale]:** On one hand, Extended RSDFO has shown to be competitive with the current state-of-the-art meta-heuristics R-PSO (even with the "extra" Riemannian logarithm map) in determining the global optimal tori. Whilst R-PSO is effective in determining the global optimal tori, it often fails to converge to the local optimal angles. At the same time, Extended RSDFO is more accurate when converging to the local optimal angles.

- **[Local scale]**: On the other hand, Extended RSDFO is also shown to be competitive with the current state-of-the-art stochastic algorithm RCMA-ES in terms of convergence to the local optimal angles within each tori, while more effective when determing the global optimal tori. RCMA-ES out-performs the other algorithms in determining the local optimal angles, but it is unable to move far away from the initial torus. That is, when given an initial point $\mu_0$ in torus number $i$, RCMA-ES can at-most explore torus numbers within $[i - 2, i + 2]$.

In this sense, Extended RSDFO gets the "best of both worlds" of R-PSO and RCMA-ES in the expense of additional computational resources. Extended RSDFO generally requires more resources (function evaluation, evaluation of boundary points, exponential maps) compared to RCMA-ES when evaluating the results across different centroids, which can be overcome by parallelization of the RSDFO streams.

Finally, RTR is inefficient in finding any optima compared to the population-based approach and is prone to be stuck in "local-local" optima.

## References


[AA19] Mazhar Ansari Ardeh. Benchmarkfcns toolbox. Retrieved July 3, 2019, from http://benchmarkfcns.xyz/, 2019.

[ABG07] P-A Absil, Christopher G Baker, and Kyle A Gallivan. Trust-region methods on Riemannian manifolds. *Foundations of Computational Mathematics*, 7(3):303–330, 2007.

[AH19] P-A Absil and S Hosseini. A collection of nonsmooth riemannian optimization problems. In *Nonsmooth Optimization and Its Applications*, pages 1–15. Springer, 2019.

[AMS04] P-A Absil, Robert Mahony, and Rodolphe Sepulchre. Riemannian geometry of grassmann manifolds with a view on algorithmic computation. *Acta Applicandae Mathematica*, 80(2):199–220, 2004.

[AMS09] P-A Absil, Robert Mahony, and Rodolphe Sepulchre. *Optimization algorithms on matrix manifolds*. Princeton University Press, 2009.

[BA10] Pierre B Borckmans and Pierre-Antoine Absil. Oriented bounding box computation using particle swarm optimization. In *ESANN*, 2010.

[BIA10] Pierre B Borckmans, Mariya Ishteva, and Pierre-Antoine Absil. A modified particle swarm optimization algorithm for the best low multilinear rank approximation of higher-order tensors. In *International Conference on Swarm Intelligence*, pages 13–23. Springer, 2010.

[BK89] Jacek Bochnak and Wojciech Kucharz. Algebraic models of smooth manifolds. *Inventiones mathematicae*, 97(3):585–611, 1989.

[BMAS14] Nicolas Boumal, Bamdev Mishra, P-A Absil, and Rodolphe Sepulchre. Manopt, a matlab toolbox for optimization on manifolds. *The Journal of Machine Learning Research*, 15(1):1455–1459, 2014.

[CFFS10] Sebastian Colutto, Florian Fruhauf, Matthias Fuchs, and Otmar Scherzer. The cma-es on riemannian manifolds to reconstruct shapes in 3-d voxel images. *IEEE Transactions on Evolutionary Computation*, 14(2):227–245, 2010.

[FKT95] Joel Feldman, H Knörrer, and Eugene Trubowitz. Infinite- genus riemann surfaces. *Canadian Mathematical Society*, 3:91–111, 1995.

[Ghy95] Étienne Ghys. Topologie des feuilles génériques. *Annals of Mathematics*, pages 387–422, 1995.

[GL12] Robert B Gramacy and Herbert KH Lee. Cases for the nugget in modeling computer experiments. *Statistics and Computing*, 22(3):713–722, 2012.

[Han06] Nikolaus Hansen. The cma evolution strategy: a comparing review. In *Towards a new evolutionary computation*, pages 75–102. Springer, 2006.

[Mil64] John Milnor. On the betti numbers of real varieties. *Proceedings of the American Mathematical Society*, 15(2):275–280, 1964.

[NW06] Jorge Nocedal and Stephen Wright. *Numerical optimization*. Springer Science & Business Media, 2006.

[Prá96] Agostino Prástaro. *Geometry of PDEs and mechanics*. World Scientific, 1996.

[PS81] Anthony Phillips and Dennis Sullivan. Geometry of leaves. *Topology*, 20(2):209–218, 1981.

[SB17] S. Surjanovic and D. Bingham. Virtual library of simulation experiments: Test functions and datasets. Retrieved July 3, 2019, from http://www.sfu.ca/~ssurjano, 2017.

[Spi79] Michael Spivak. A comprehensive introduction to differential geometry, publish or perish, 1979.

[TAV13] Roberto Tron, Bijan Afsari, and René Vidal. Riemannian consensus for manifolds with bounded curvature. *IEEE Transactions on Automatic Control*, 58(4):921–934, 2013.

[Tog73] Alberto Tognoli. Su una congettura di nash. *Annali della Scuola Normale Superiore di Pisa-Classe di Scienze*, 27(1):167–185, 1973.

[Whi44a] Hassler Whitney. The self-intersections of a smooth n-manifold in 2n-space. *Annals of Math*, 45(220-446):180, 1944.

[Whi44b] Hassler Whitney. The singularities of a smooth n-manifold in (2n- 1)-space. *Ann. of Math*, 45(2):247–293, 1944.

# Chapter 10
# Conclusion and Future Research

**Abstract** Stochastic optimization on Riemannian manifolds is a topic scarcely explored. At the writing of this book, the field is still in its infancy. In this last chapter of the book, the materials discussed throughout the book are wrapped up in a gentle fashion and possible directions for future research are outlined.

## 10.1 Conclusion

Manifold optimization is an emerging field of contemporary optimization, motivated partly by the growing interest in geometry within information theory. At the time of writing this book, applications that necessitate manifold optimization techniques are scarce. Manifold optimization methods developed in the literature thus far revolve around compact manifolds with global canonical vectorial representations, examples include $n$-spheres in sparse PCA [AH19], and matrix manifolds in low-rank matrix completion [CA16] and manifold-valued image processing [BBSW16]. These in turn depend strongly on the specific structures of the search space manifolds, rather than the general manifold structures.

On the other hand, whilst classical optimization methods in Euclidean spaces have been well-studied over the past centuries, these techniques cannot be simply translated to the context of Riemannian manifolds. Researchers who are more accustomed to classical optimization techniques may therefore be more inclined to approach the compact manifolds in the literature with readily available constraint optimization techniques instead, and as a result the structure of the underlying manifold search space is ignored.

The work of this book advanced along these two directions. We developed a geometrical framework for population-based stochastic optimization on Riemannian manifolds, where components of the algorithms can be studied from a geometrical perspective. To motivate and necessitate manifold optimization, we presented a synthetic example of Jacob's ladder in this book, a surface of potentially infinite genus that cannot be solved with classical constraint optimization techniques. In this final



R. S. Fong and P. Tino, *Population-Based Optimization on Riemannian Manifolds*, Studies in Computational Intelligence 1046,
https://doi.org/10.1007/978-3-031-04293-5_10

chapter we summarize our work on population-based stochastic derivative-free optimization methods on general Riemannian manifolds, and outline possible directions for future research.

Inspired by the geometrization of statistical models in Information Geometry and the recent developments of manifold optimization, this book aimed to provide a geometrical interpretation of population-based stochastic optimization over Riemannian manifolds. The aim was to construct a geometrical framework where the intricacies of stochastic optimization algorithms over manifolds can be studied using the statistical geometry of the decision space and the Riemannian geometry of the search space.

To this end, we took the long route and investigated information geometrical structures of statistical models over manifolds in Part I.

We began our investigation by laying the foundations, and described the essential elements of two disciplines of geometry in the literature: Differential Geometry and Information Geometry. The former establishes the geometrical structure of the search space, while the latter describes the geometry of the decision space.

We then examined two ways to construct intrinsic statistical models on Riemannian manifolds in the literature, which can be classified roughly as a "statistical" approach and a "geometrical" approach. We discussed how neither approach is suitable for our desired geometrical framework: the "statistical" approach is too restrictive, as the construction is bounded by the normal neighbourhoods. Whereas the "geometrical" approach is too general, as the relationship between statistical estimation and the parameters of the information geometric structure is elusive.

Using the machineries from Differential Geometry and Information Geometry, we combine the essence of the two approaches. In particular, we generalized the use of Riemannian exponential map in the "statistical" approach to a local orientation-preserving diffeomorphic bundle morphism on the base manifold $M$. By viewing the locally inherited probability densities on $M$ as probability $n$-forms, we showed that a dualistic statistical geometry on family of local densities can be inherited entirely from a family of probability densities on Euclidean spaces.

The construction of locally inherited densities and its corresponding statistical geometry is then extended beyond the confines of normal neighbouhoods to totally bounded subsets on $M$. We described a family of mixture densities on totally bounded subsets of $M$ as a mixture of the locally inherited densities, and its product statistical geometry is derived therein.

Equipped with the geometrical framework from the Part I, we return to our original investigation of optimization on Riemannian manifolds in Part II.

We began our investigation by reviewing manifold optimization in the literature, and we observed that the adaptation of optimization algorithms from Euclidean spaces to Riemannian manifolds follows the same principle as the "statistical" approach of locally inherited probability densities. That is, the locality from the "statistical" approach still persisted, and the operations of the Riemannian adapted algorithms are once again confined by the normal neighbourhood. Furthermore, whilst some stochastic algorithms on Euclidean space admit an information geometrical interpretation, the Riemannian counterpart under the adaptation framework in the literature does not share the same property.

These issues are accentuated by a generalized framework for adapting Stochastic Derivative-Free Optimiztion (SDFO) from Euclidean spaces to Riemannian manifolds. Under the notion of locally inherited probability densities described in Part I, RSDFO partially addresses the issues by providing a local information geometric interpretation of Riemannian adapted SDFO. However, the local restriction still lingered.

In order to overcome the local restrictions of RSDFO, we required a parametrized probability densities defined beyond the normal neighbourhoods of Riemannian manifolds, and the mixture densities described in Part I provides exactly what is needed. The product Riemannian structure of parametrized mixture densities on Riemannian manifolds thus gave rise to a population-based stochastic meta-algorithm—Extended RSDFO based on the foundation of RSDFO. We discussed the geometry and dynamics of the evolutionary steps of Extended RSDFO using a modified Fisher metric on the simplex of mixture coefficients, and showed that Extended RSDFO converges globally eventually in finitely many steps on connected compact Riemannian manifolds.

Finally, we wrapped up our investigation by comparing Extended RSDFO with state-of-the-art manifold optimization methods in the literature, such as Riemannian Trust-Region method [ABG07, AMS09], Riemannian CMA-ES [CFFS10] and Riemannian Particle Swarm Optimization [BIA10, BA10], using optimization problems defined on the $n$-sphere, Grassmannian manifolds, and Jacob's ladder.

From the experimental results, in particular the ones on Jacob's ladder, we observed that model-based approaches and evolutionary strategies perform better than gradient-based methods in mutli-modal manifold optimization problems. Even though Extended RSDFO is implemented with a basic RSDFO as the "core" algorithm in the experiments, we demonstrated that Extended RSDFO is comparable to R-PSO (meta-heuristic) in finding global optimal, whilst also comparable to the accuracy of RCMA-ES (model-based algorithm) in terms of attaining the local optima. In this sense, Extended RSDFO has the "best of both worlds" scenario at the expense of possibly additional computational resources.

## 10.2 Future Research

Stochastic optimization on Riemannian manifolds is a topic scarcely explored. To the best of our knowledge this is the first monograph that presents both theoretical and empirical aspects of stochastic population-based optimization on abstract Riemannian manifolds. Therefore, the work discussed in this book is still in its infancy. For the remainder of the book, we wrap up our discussions by outlining possible directions for future research.

- **[Natural parameters and application of mixture densities]**:
  In the first part of the book, we developed a family of mixture parametrized densities over totally bounded subsets of Riemannian manifolds $M$, and derived its product statistical geometry.
  Under this framework, the local statistical point estimations are consistent with the statistical parameters of the local component distributions within the normal neighbourhoods. It is natural to ask whether it is possible to infer from the parametrization a notion of "globally" defined statistical object on $M$.

- **[Components of Extended RSDFO]**
  Extended RSDFO described in Chap. 8 can be implemented with a variety of Riemannian SDFO "core" algorithms. From the experiments described in Chap. 9, we observed that Extended RSDFO implemented with a very basic RSDFO "core" is comparable to both R-PSO and RCMA-ES on both the global and local fronts. Further research is thus required to study the impact of using a different "core" algorithm in Extended RSDFO. Moreover, more complex selection schemes for search centroid selection can be incorporated in Extended RSDFO to enhance the diversity of solutions.

- **[Parallelization of Extended RSDFO]**
  One of the advantages of the product statistical geometry of mixture densities lies in the independence of parametrization of mixture components. This allows us to compute the evolution and expected fitness of the local RSDFO modules of Extended RSDFO separately and independently. Parallelization of the RSDFO core local modules would provide more efficient implementations of Extended RSDFO compared to the sequential version of Extended RSDFO described in Chap. 9.

- **[Further study on Convergence Analysis of Extended RSDFO]**
  The current convergence studies are based on the metric space structure of $M$ under the Riemannian metric topology. We showed that Extended RSDFO converges eventually globally within finitely many steps in compact connected Riemannian manifolds $M$, under the sufficient condition that Extended RSDFO generates a boundary exploration point for each iteration.
  Further investigation is needed to derive more explicit relations between the manifold structure of the search space, the information geometric structure of the decision space, and complexity of Extended RSDFO.

# References


[ABG07] P-A Absil, Christopher G Baker, and Kyle A Gallivan. Trust-region methods on Riemannian manifolds. *Foundations of Computational Mathematics*, 7(3):303–330, 2007.

[AH19] P-A Absil and S Hosseini. A collection of nonsmooth riemannian optimization problems. In *Nonsmooth Optimization and Its Applications*, pages 1–15. Springer, 2019.

[AMS09] P-A Absil, Robert Mahony, and Rodolphe Sepulchre. *Optimization algorithms on matrix manifolds*. Princeton University Press, 2009.

[BA10] Pierre B Borckmans and Pierre-Antoine Absil. Oriented bounding box computation using particle swarm optimization. In *ESANN*, 2010.

[BBSW16] Miroslav Bacák, Ronny Bergmann, Gabriele Steidl, and Andreas Weinmann. A second order nonsmooth variational model for restoring manifold-valued images. *SIAM Journal on Scientific Computing*, 38(1):A567–A597, 2016.

[BIA10] Pierre B Borckmans, Mariya Ishteva, and Pierre-Antoine Absil. A modified particle swarm optimization algorithm for the best low multilinear rank approximation of higher-order tensors. In *International Conference on Swarm Intelligence*, pages 13–23. Springer, 2010.

[CA16] Léopold Cambier and P-A Absil. Robust low-rank matrix completion by riemannian optimization. *SIAM Journal on Scientific Computing*, 38(5):S440–S460, 2016.

[CFFS10] Sebastian Colutto, Florian Fruhauf, Matthias Fuchs, and Otmar Scherzer. The cma-es on riemannian manifolds to reconstruct shapes in 3-d voxel images. *IEEE Transactions on Evolutionary Computation*, 14(2):227–245, 2010.

# Index





R. S. Fong and P. Tino, *Population-Based Optimization on Riemannian Manifolds*,
Studies in Computational Intelligence 1046,
https://doi.org/10.1007/978-3-031-04293-5

www.ingramcontent.com/pod-product-compliance
Ingram Content Group UK Ltd.
Pitfield, Milton Keynes, MK11 3LW, UK
UKHW021834270726
14058UKWH00001B/145
*9783031042959*